THE GREAT FOOD GAMBLE

Also by John Humphrys

Devil's Advocate

JOHN HUMPHRYS

The Great Food Gamble

Hodder & Stoughton

First published in Great Britain in 2001
by Hodder and Stoughton
A division of Hodder Headline

1 3 5 7 9 10 8 6 4 2

A CIP catalogue record for this title is available from the British Library

ISBN 0 340 770 45 7

Typeset in Linotype Trump Medieval by
Rowland Phototypesetting Ltd,
Bury St Edmunds, Suffolk
Printed and bound in Great Britain by
Clays Ltd, St Ives plc

Hodder and Stoughton
A division of Hodder Headline Ltd
338 Euston Road
London NW1 3BH

For Owen
who loves his food

CONTENTS

ACKNOWLEDGEMENTS

I would double the length of this book if I listed all the ways in which so many people have helped with my research, but let me list a few without whom it would not have been possible: Dr Vyvyan Howard, Janie Axelrad, Professor Hugh Pennington, Dr Sue Mayer, Dr Erik Millstone, Peter Beaumont, Dick Thompson, Professor Gordon McVie, Dr Bob Masterton, Richard Young, David Wilson, Mark Purdey, Don Stanniford, Tony Coombes, Dr Vivienne Nathanson, Sandra Bell, Dr Richard Fortey, Dr David Barling, Professor Phil Ineson, Lizzie Vann, Peter Seggar, Patrick Holden and Simon Adey Davies.

Ruth Evans helped me pull everything together and lit a candle on those dark nights of the soul that afflict everyone who ever attempts a book.

I am grateful to them all.

FOREWORD

I started thinking about this book in the early nineties when the full horror of BSE was becoming apparent. Many people were beginning to ask serious questions about the food on our tables and to worry about food in a way we had not done in our recent history. I thought then that we needed a proper national debate which addressed the most fundamental questions. Sadly, it took another great farming crisis to bring that debate about and it is unfolding as I write these words in March 2001.

That crisis is, of course, the foot-and-mouth epidemic. The question on which it has focused attention – and which should have been addressed a long time ago – goes to the heart of Britain's food policy since the Second World War. Has the relentless drive for more and more cheap food proved a mistake?

The most powerful voices in the food industry scoff at even a hint of doubt. They tell us it is naïve even to raise the question. They remind us that choice is more varied and shopping for food has never been more convenient. Above all, they say, food is cheaper than it has ever been. The only alternative to the way we have been farming, producing and distributing our food for half a century is to return to some primitive form of agriculture and food production that would lead to desperate shortages, sky-high prices and disease-ridden animals. We would be back in the Middle Ages. Some talk darkly of malnourished children, even the return of diseases such as rickets.

Most of that is hysterical, self-serving nonsense. Malnutrition is a function of poverty or ignorance or both, not of the availability of food. I have spoken to doctors who worry about some of their young patients because they are fed a diet of junk food. Their parents have never cooked them a meal of fresh meat and vegetables in their short lives. That has nothing to do with the price of food. On the contrary, processed food is invariably more expensive than fresh vegetables. It is, to use the jargon of the day, a 'lifestyle' choice and one that is encouraged by the industry. Quite simply, there is more profit in a bag of crisps than in a pound of potatoes.

Nor is a return to 'primitive' farming practices the only alternative to factory farming and highly intensive agriculture. That is a gross insult to all those farmers who care deeply for the welfare of their animals and do not regard them as mere units of production. It also ignores the advances made in less intensive farming technologies. A growing number of farmers are finding ways of achieving good yields without tearing open yet another drum of chemicals or bag of synthetic fertiliser.

But the big question the industry and most politicians have been so reluctant to address is whether 'cheap' food is really cheap. To do so is to raise doubts about their judgement or even their motives. What price, for instance, should society put on the destruction of so much of our rural heritage, the loss of our water meadows and ancient hedges, the disappearance of so many songbirds?

It may be impossible to calculate that sort of thing in hard cash, but much else can be quantified. There are the taxes we pay to finance farming subsidies. Many of them go to some of the wealthiest farmers in the land to grow food we do not need – or even to pay them not to grow it. There is the cost of cleaning chemical pollution from our drinking water. There are the consequences for the National Health Service of factory farmers abusing antibiotics. There is the possible impact on our health of chemical residues in our food.

There are the long-term effects of soil erosion and declining soil fertility. There is the terrible impact and vast cost of a tragedy such as BSE. And now, as I write, we are in the midst of another epidemic, foot-and-mouth disease. It would not be fair to say it is the direct result of intensive agriculture. But modern practices of food production and supply have enabled it to spread at a terrifying speed across the entire country.

Is it naïve to raise questions about a food policy that has created such a legacy? I think not. That is the reason I have written this book. Some of the things I have learned during the course of researching it have worried me a great deal. But this is not a counsel of despair. There are many farmers and food producers and even politicians who accept that mistakes have been made and are searching for better ways of doing things. I believe the British people will insist on it. We are no longer prepared to take our food for granted. The nation wants a serious debate and there is no way of stopping it.

DRIVEN BY NEED

Tom Connelly was a relieved man. He had managed to book a passage home from Liverpool to Montreal on the passenger liner *Athenia* for himself, his wife and their three sons. The *Athenia* was not the most luxurious liner afloat. She was sixteen years old and showing her age a little. Yet every berth in every cabin was booked and there were so many people wanting to make the voyage to Canada that they could all have been sold several times over. Like Connelly, they were fleeing from a country about to go to war with Germany. The date was Saturday 2 September 1939.

Neither the Connelly family nor any of the other 1,418 passengers and crew had any reason to worry as the *Athenia* slipped anchor in the late afternoon sunshine. Even if the announcement of war were to be made while they were at sea they had the reassurance of knowing that their ship was protected by international law from attack by enemy submarines. Under pressure from Great Britain, Germany had signed the 1936 Submarine Protocol restricting its commanders to operating against merchant shipping under the Hague Convention Prize Laws. Those laws made it illegal to sink a ship without having first placed passengers and crew in a place of safety. So they might have lost their possessions, but not their lives. And anyway their captain, James Cook, had told them that he was taking no chances. He was increasing speed and sailing a different route, pursuing a zigzag course.

The following morning, as the Connelly family was finishing its first breakfast on the *Athenia*, the British ambassador in Berlin was on his way to the Chancellery building with a telegram for Adolf Hitler. It contained an ultimatum. If the German invasion of Poland, which had begun on 1 September, was not halted immediately and an assurance given that the German troops would be withdrawn, then Britain would declare war on Germany. Two hours later the emotionless voice of the British Prime Minister, Neville Chamberlain, was broadcast across the airwaves of the BBC.

'I have to tell you,' he announced, 'that no such assurance has been received and that consequently this country is at war with Germany.'

Even as Chamberlain spoke, more than a third of Germany's small fleet of U-boats was on patrol in the waters of the North Atlantic. Their commanders could have been in no doubt about their orders. At two o'clock on that Sunday afternoon they had received a signal from U-boat Command reiterating what they already knew: 'All passenger liners carrying passengers are allowed to proceed in safety. These vessels are immune from attack even in convoy.' But the clearest rules have no force for those who choose not to obey.

U30 was under the command of an ambitious twenty-six-year-old, Lieutenant Fritz-Julius Lemp, and it was the tragedy of the *Athenia* that he should have been closest to her course. In the early evening of 3 September the ship was heading into the Atlantic swell off the northernmost tip of Ireland when her single funnel and black-painted hull was spotted from the conning tower of U30. Later Lemp was to claim that he had identified her as an armed merchant cruiser. She would indeed have been a legitimate target if there had been the smallest sign of any weapons on her decks. There was none. The other explanation was that Lemp wanted to make history by bagging the first target of the war. In that, he succeeded. Tom Connelly was back in his state room when the first torpedo struck. His youngest child was in bed.

'Without being told we realised what had happened,' he told reporters later. 'We only waited to throw some clothes on the child and then made a dash for the lifeboats. One of the hatches had been blown right up through the deck and many passengers were badly injured by flying splinters. My wife had a deep gash in her forehead which bled profusely but we had no time in the rush of the moment to attend to her injury . . . We were fortunate in keeping our family group together and getting into the same lifeboat, which seemed to be crowded to an alarming extent. There were only seven men in the boat and we had to man the oars. We realised it was essential to keep moving.'

They spent ten hours in the small boat and were dropping with exhaustion when the rescue ships, alerted by the signal for U-boat attack – 'SSS . . . SSS . . . SSS' – began arriving. Twenty-four hours later the Connelly family was back in Britain, the *Athenia* lay at the bottom of the sea and 118 men, women and children were dead.

The sinking of the *Athenia* sent a powerful signal to the people of Britain. Whatever the conventions of war may have said, no ship was safe from the guns and torpedoes of German warships. If they could sink a passenger liner they could and would do the same to any cargo ship that came into range. In that first week of the war the U-boat fleet consisted of only fifty-seven vessels, but the shipyards of Germany were about to start building more. Many more. For an island nation facing a battle for its very survival it was a threat that could not be ignored. Britain was heavily dependent on food bought from abroad. Shipping was its lifeline. If the U-boats could not be stopped, then the nation's dependence on imported food had to be lessened.

In the days and months that followed the loss of the *Athenia* that lesson was rammed home with every German torpedo that found its target. Not all U-boat commanders showed the same callous disregard for innocent life displayed by Lieutenant Lemp. When the British freighter *Olivegrove* spotted a stalking submarine her skipper, Captain Barnetson, tried to escape. But his old tub was no match for the sleek submarine and a warning shot was fired. He surrendered to the U-boat commander, expecting to be taken prisoner at best. Instead he and his crew were put on board their lifeboats. The German officer made sure their navigational equipment was working properly and then told them he would put them on a straight line for the Irish coast.

'We followed him for some hours until he sent up two red rockets and told us there was a steamer ahead and it would rescue us,' said Captain Barnetson. One member of the crew even managed to take his pet canary with him.

But – honourable though he may have been – the German officer still sent the *Olivegrove* to the bottom of the Atlantic and, with her, thousands of tons of desperately needed food. When war was declared Britain had scandalously little food stocks stored away: only three weeks' supply of wheat and one month's of sugar. The prospect of a nation of tea-drinkers being deprived of its sugar sent shivers down the spines of every civil servant in Whitehall and every politician across the square in Westminster. German bombers were one thing; going without a cuppa was quite another. The Ministry of Food accepted that 'we have reached conditions of dangerous scarcity'.

Goebbels, the German propaganda minister, recognised an opportunity when he saw it and had Rudyard Kipling's poem, 'Big Steamer', revised so that it read:

> For the bread that you eat and the biscuits you nibble,
> The sweets that you suck and the joints that you carve.
> They are brought to you daily by all us big steamers,
> And if anyone hinders our way then you'll starve.

The country could no longer rely on imports. Where the *Athenia* and the *Olivegrove* had led, countless more ships – big and small – were to follow. At the start of the war Britain had been importing more than 22 million tons of food and animal feed, much of it carried by a fleet of 3,000 merchant ships across the North Atlantic. Within less than three years that had fallen by more than half. Something had to replace it – and the nation had to change its eating habits. Before very long every home in Britain had its own little buff book. Rationing had arrived.

The Ministry of Food devised an emergency basic diet: 12 ounces of bread, 2 ounces of oatmeal, 1 pound of potatoes, 1 ounce of fats, 6 ounces of vegetables and six-tenths of a pint of milk. Not bad for a supermodel wanting to keep her figure trim and a damn sight healthier than most modern diets; you certainly wouldn't starve on it. But there had to be more to life than bread and oatmeal and whereas the nation had been happy to tighten its belt in wartime we expected more on our plates in peace. We also wanted to be reassured that we would be less vulnerable if we were ever again threatened in the same way. So when the government produced a new food policy based on maximum production, postwar Britain applauded. The more grain we could grow, the more pigs we could fatten, the more milk we could produce, the better for all of us. If it meant changing farming practices and also changing the face of the British countryside, then so be it. And it worked. From the 1950s production rose as never before and the price of food began to fall.

But then other factors began to come into play. Over the decades the memories of the war began to recede. We realised that if, God forbid, there were to be another one it would be a very different kind of war. The threat would come from unseen missiles bearing nuclear warheads rather than unseen submarines with their relatively puny torpedoes. It would all be over in a series of terrible flashes before we'd had time to think about stocks of grain running out.

We were also beginning to realise something else. It was undeniably true that we were paying less for our food at the tills and the supermarket checkouts, but slowly it dawned on some that there were other prices to be paid for intensive agriculture. The farmers were being bribed to produce ever more food through an elaborate system of subsidies. Every taxpayer in the land bore the cost of those subsidies. Viewed from that perspective the cheap bread and milk seemed rather more expensive.

There were other costs, too. As the farmers were being bribed, so the land was being forced. Even the richest soil cannot go on and on producing bumper harvests of grain, year in and year out, without being allowed to recover its fertility. The old way had been to allow fields to lie fallow or to rotate the crops with something like clover so that the soil had a chance to regain its fertility. The new way was to open a sack of chemical fertiliser and spread it on the land. Then to open a drum of herbicides to kill the weeds. Then to open yet another drum of pesticides to kill the insects. However hungry the crops and however careful the sprayers some of the chemicals inevitably found their way into the water courses, sometimes a great deal. They poisoned the ditches that fed the streams and the rivers and seeped into the water table that fed our homes. The poisons had to be cleaned out of the water and the cost of the clean-up was great.

There was yet another cost – one that could not be calculated in pounds and pence – because some of the chemicals did their work only too well. So well that even the least productive land could be coaxed into delivering its bounty of wheat or barley or lamb. Millions of acres of downland and moorland that had been gently grazed for centuries now went under the plough. Water meadows and wetlands were drained.

Ancient hedges and copses were ripped out. In England alone hedgerows disappeared at the rate of more than 10,000 miles a year – far more at the height of the 'boom'. The environment paid a fearful price for those overflowing silos and even, at the height of the madness, disused aircraft hangers crammed with unwanted grain.

And then there was the final cost: the threat to the quality and even the safety of the food itself. As one scare followed another we grew increasingly uneasy about the food on our plates. When mad cow disease struck it was the final straw. It destroyed whatever vestige of confidence remained. The financial cost was staggering – approaching five billion pounds. The cost in human suffering was incalculable: young lives lost to the human version of the disease, variant CJD and the real possibility of many more to come.

Over those years of ever growing harvests few people bothered to express any doubts they may have had about the underlying agricultural policy. The only god to be worshipped was the god of maximum production and non-believers brave enough to question it were met with ridicule and derision. The vested interests that lay behind intensive agriculture were all-powerful: the mighty agro-industrial companies, the National Farmers' Union, the so-called barley barons, governments packed with landowners whose bank accounts grew fatter as their grain silos bulged with grotesquely subsidised corn. Their arguments in defence of the policy that had suited them so well for so long never varied. They went something like this.

No matter that public confidence was being rocked by one food scare after another or that Britain's glorious rural heritage was being destroyed before our eyes, it was obvious to the meanest intelligence that there could be no alternative. Did these naïve dreamers not realise that if we went back to the mythical 'good old days' before efficiency ruled the land we would all starve by nightfall? And just take a look around. Only a fool could fail to see how our children have grown big and strong and healthy, how our life expectancy has increased so much that our biggest worry is paying the pensions of an ageing population. Those poor idiots with their beards and sandals, their belief in muck and magic, were harmless enough as entertaining diversions but God help us if we were ever to take them seriously.

It proved a powerful defence – until the closing years of the twentieth century. By then the beard and sandal brigade had shaved off their beards and cast off their sandals. Their muck and magic had been replaced by solid research and trial and error. Their harvests may not have matched those of the barley barons, but they delivered some impressive results in the toughest laboratory of them all – the field and the market-place. Yet there was still that other argument to deal with: the health of the nation.

True, we live longer and no one can quarrel with that. But might not better medicine and effective vaccines have had something to do with that? Diseases such as diphtheria and TB, whooping cough and polio, which once filled the surgeries and the graveyards, have been defeated. Yes, our children grow ever bigger and ever stronger and mature ever earlier. But do we really want little girls menstruating at the age of ten?

So today, more than half a century on, the certainties of those postwar years and the food policies that followed seem less certain. Yet still the powerful vested interests and – with one or two exceptions – our political masters refuse to accept that we may have got it wrong. Not for the first time in a democracy's history, it is ordinary people who are making the running: telling the politicians that they've had enough of food scares and environmental disasters; telling the supermarkets that if they want genetically modified food they will make the decision for themselves.

A growing number of people believe that the time has come to pause, to look at other ways of producing the food that we need. To listen to other voices than those of the agrochemical industry and the big landowners. To acknowledge that there are serious questions as to whether the system that was born out of the best possible motives is sustainable for very much longer, not only in terms of the environment but in terms of what it may be doing to our health. All life is about taking risks, of accepting gambles of one kind and another. But this is the biggest gamble of them all. We should weigh the odds very, very carefully.

YESTERDAY, TODAY . . .
AND TOMORROW?

March 1950

A mixed farm on the outskirts of a market town somewhere in England

The eight-acre field looks as though it has been given a haircut by a barber after a very long liquid lunch. The intention was clearly to create a fifties crew-cut but it hasn't quite worked out. Parts of it are entirely bare and the brown earth shines through where the clippers slid too close. At irregular intervals longish tufts stick out – missed entirely by the drunken barber. Most of the field is stubble, which is what it is meant to be: the remnants of a crop of barley that stood tall and golden six months ago, before the combine harvester clattered across its lumpy surface. To the casual human observer the field may seem dead, the stubble rotting away and the soil as lifeless as the lightning-scorched oak tree that stands bare-limbed and blackened in one corner. But to the small mammals, birds, insects, bacteria, and the billions of micro-organisms that call it home, this is Benidorm in August, Trafalgar Square on New Year's Eve. From the perspective of a micro-organism this is where it all happens.

It has been happening for a long time. Long before the first human discovered he could walk upright, long before the first dinosaur cracked its wobbly way out of the first egg, these tiny beasts have been beavering away. Some of them arrived on dry land and began to establish their complex ecosystems four hundred million years ago. They've had a great deal to put up with since then. They not only managed to cope with the ice age; they also survived the first great global cataclysm that wiped out ninety per cent of everything on earth and then the second one, sixty-five millions years ago, that saw off the dinosaurs. We should be grateful for that. If they had not survived, we would never have existed.

Their job is to recycle nutrients. The equation is a simple one: no recycling equals no life. Every living thing in this eight-acre field is, ultimately, recycled. From the stubble of the barley crop harvested last autumn to the first leaves on the tiny oak sapling when it first struggled for life in the corner of the field two centuries ago to the great leathery hides and massive bones of the mastodons who once thundered across these plains . . . it is all broken down into minute particles, its energy and its nutrients released so that something else can grow in its turn.

If that cycle ever stops there will be no more sun-filled romps for us in Benidorm, no more rowdy drunks counting down to a happy new year in Trafalgar Square. In fact, there will be no more new years full stop – happy or otherwise.

If all this activity beneath the surface of the earth is ever seriously threatened then we all pay a price. Ultimately life on earth depends on vegetable productivity. Without it life will become impossible. And it sometimes seems that we're doing our damnedest to stop it. If we compressed our time on earth into one twenty-four-hour period the past fifty years would register as a micro-second. We have done more to disrupt the cycle in that micro-second, that blink of an eye, than in our entire history.

But this spring day in 1950 in this particular field in England is a good time and a good place to be a micro-organism. The two small boys who climb the gate into the eight-acre field neither know nor care about the activity beneath their wellington boots, but they know the field well. It is a short walk from their cottage in the nearby market town and a favourite haunt. One of the branches of the old oak supports a car tyre hanging from a rope, a riskier but far more interesting swing than the ones on offer in the local park. When the barley grass is young they wedge it between thumb and index finger and make hideous noises with their lips. When the barley is ripening they steal the corn to chew it. There is almost always something to pick from the hedges. In the early spring their mother sends them to nip the fiddle-shaped tops from the young fern plants, which she fries in butter for tea. Or to pick the wood sorrel leaves to mix with the salad. Or fresh young stinging nettles to be made into soup. Their favourite food is in the autumn: blackberries. If they manage to get any from hedge to home they will end up in pies with some windfall apples; usually they end up in their stomachs long before they reach the pot. Now, they are looking for eggs to add to their collections.

In the bigger, rougher hedges are the nests of bullfinches, white throats and even barn owls. Linnets and corn buntings are everywhere, feasting off the seed scattered on the ground after the inefficient harvester has done its work. So are hedge sparrows, song thrushes, blackbirds and yellowhammers, with the extraordinary song that they mimic: 'a little bit of bread and no cheese'. There are lapwings, too, in the wetter parts of the field. They have yet to be protected by the Wild Bird Protection Act of 1953, which will finally put a stop to people tucking into them with their Sunday lunch.

When the boys are a little older they will find part-time work on the farm, helping to pile up the bales of hay and straw, weeding the few rows of vegetables and picking apples from the small orchard. Their father works here full time. It was he who harvested the oats last autumn. Once it has been dried it will be stored until it is needed and crushed, to be fed to the cows when they come in for milking. Nothing is wasted here. The straw left behind in the field will dry in the sun and then it will be baled, useful for bedding in the cow sheds in winter and palatable enough for the cows to eat. It's good for them too: it helps make the milk they produce a little creamier. In winter their main diet is hay, cut when the grass has grown long and begun flowering. If the weather is kind and the sun shines for days on the mown grass the hay will smell sweet and the cows and sheep will love it. But hay-making is a risky business. Too often the rain falls just as the hay is almost ready to be baled, and then the drying has to start all over again. The farmer dreads those terrible years when the hay never dries. It's too risky to bale it when it is damp. He recalls with a shudder what happened to one of his neighbours a few years ago.

It had been a miserable, wet summer. The grass had grown well enough and there was plenty of it when the mower sliced through it on a promisingly sunny day in late June. But the promise faded as the clouds gathered. It rained that night and most of the next day. Then the sun tried again and, with it, a strong, drying breeze rippled through the limp grass. Twice the farmer climbed on his tractor and rode up and down his fields, turning the grass to let the sun and breeze do its work more effectively. The next morning he started preparing the baler to parcel the hay, optimistic that at last he would be able to bring in a decent harvest to feed the cows through the long winter. And then the heavens opened again.

It rained just long enough to soak the grass through before the sun made a watery return. Two days later the farmer, knowing the hay was not as dry as it should be, decided to make the best of a bad job and baled it up. It was a fatal mistake that would cost him dear.

From the moment the hay was packed tight into his barns it began to heat. The farmer, busy with a hundred other tasks and thankful at least to have the hay harvest in, knew nothing about it. He was sound asleep weeks later when the dogs woke him with their panicked barking. He smelled the smoke before he saw the flames. The main barn was ablaze. The damp hay had built up a fearsome heat, ignited spontaneously and was now burning down the barn. By the time the firemen arrived it had taken the other two barns with it and every last bale of hay. He was not insured and, since he already owed the bank more money than he could really manage, it was enough to finish him off.

In the very worst years the hay never dries enough even to think of harvesting it. Instead the rain reduces it to a slowly blackening mess, good for nothing except to be ploughed into the ground for the subterranean army of insects and micro-organisms to turn it back into soil. So the owner of the eight-acre field is thinking of trying something new that he has heard works well. He plans to build a silage clamp. Then the grass will be mown when it is still young and packed with green juices. Instead of being left for days to dry in the field it will be chopped into small pieces, allowed to wilt in the field for a day, then dumped in the concrete clamp, rolled over and over again to squeeze out the air and covered with black plastic sheeting. Then it will be left to ferment and when winter comes and the cows are brought in from the fields . . . delicious. Or so the cows seem to think. It's not foolproof – even with silage the weather plays a big part – but it's much less risky than haymaking and if the silage is good it is even more nutritious than the hay.

What matters to this farmer in the spring of 1950, and to thousands more like him throughout the country, is that he should be able to grow almost everything he needs and make use of everything. The pigs get the apples that fall from the trees and the chickens get the scraps from the table and the sweepings from the grain sheds. The land gets the dung from the cows and pigs and sheep. Even the air itself is used: the clover planted with the grass seed takes nitrogen from the atmosphere and passes it through its roots into the soil, leaving it richer than it found it. This is what good farming and good husbandry has been about for more than ten thousand years, ever since we stopped relying on wild animals and berries.

It is about maintaining the land in the best possible condition, rotating the fields so that a crop that needs plenty of nitrogen is planted in a field where clover grew the year before. It is about selecting varieties of barley or wheat that grow so tall they deny light to the weeds and kill them off. It is about maintaining hedges for the birds and small mammals to live in, knowing that they will pay their rent by feeding off the insects which would otherwise feed off the crops. It is about treating the animals with consideration, not trying to force too much milk from the cows or make the beef grow too fat too quickly.

When the farmer ploughs his fields in the spring after the winter frost has done its work the rich soil falls from the plough share easily and the birds follow the plough. There are many meals to be made from the worms but there will still be plenty left when they and their chicks have had their share.

The farm will never make its owner rich but it will provide a decent living for him, his family and several men in the nearby village, including the father of the boys playing in the eight-acre field. There's an old saying in farming that the difference between success and failure is two weeks. Every farmer knows what that means. Do the planting and the harvesting at the right time, judge the weather and the soil conditions reasonably accurately, and you will reap the benefits. Get it wrong and it will be a lean year. Most years the farmer gets it more or less right, but because it is a mixed farm and does not rely on one crop it is not a disaster if something lets him down.

The wheat that he grows on his richest, free-draining land usually makes a decent profit, but not always. That, too, depends on the weather. If the price of milk from his dairy herd falls, as it has done from time to time over the years, then he may make a better profit from selling his pigs for bacon. If pork is unprofitable one year, the lambs from his small flock of sheep may sell for a better price in the local market. And then there are the male calves born to his dairy cows. The females will grow to take their place in the herd in three years' time but the bull calves will be fattened up on grass and a little barley and end up as the Sunday roast. The old cows that can no longer produce enough milk to justify their keep will end up on the dining table, too: not at smart dinner parties as prime cuts of fillet steak, but as stewing meat for school dinners. The apples are a useful little earner and so are the chickens that scratch around the farmyard. They produce only a very modest income but, more importantly, keep his growing family in fresh eggs.

The owner is farming much as his father did a generation ago and his grandfather before him. The old shire horses that once pulled the plough and the mower have been retired long since and the tractor has taken their place. When the corn is harvested another machine will be called into service. Otherwise, the farming practices are virtually unchanged. A medieval farmer transplanted to this time would be puzzled by the machinery, but by little else. The farmer is relying on mostly natural forces to produce and protect his crops. Apart from diesel oil and machine parts, he buys in virtually nothing. He is farming in harmony with nature. He believes that so long as his soil is in good heart and the weather is not too unkind he can produce the food that is needed. That is how it has always been and he can see no reason why it should change. One day he expects the farm to pass to his eldest son and then to his grandchildren. He views the future with equanimity.

There are many farms like this in Britain in this spring of 1950, some bigger but most smaller. More than a million families earn their living from 450,000 mixed farms. The market town in which the two small boys live serves the farms in the same way that the farms serve the town. At the weekly cattle market sharp-eyed men make silent bids for curious young calves and nervous heifers with nods that are seen and understood only by the auctioneer. A stranger who looks for some satisfaction or disappointment in the face of the farmer selling or the dealer buying will not find it. There is no place for emotion in the auction ring. The stall holders in the food markets make up for their silent colleagues, loudly praising the freshness of their vegetables, the quality of their cheeses.

The mother of the two boys comes here every week. Because she has no fridge, she shops almost every day. Her milk is delivered to the doorstep, creamy and unpasteurised, and at weekends she separates the cream for Sunday tea. For a few glorious weeks in summer there are fresh strawberries to pour it over but at this time of year the choice of fruit is limited and what there is can be too expensive for her budget. But the root vegetables and green vegetables are usually good and relatively cheap and in a few weeks the spring lamb will be in the butcher's. By the autumn she will be able to pick and choose between a dozen varieties of apples and pears, plums and berries. Some will be preserved for the winter months when the only alternative is tinned fruit. If she thought about it she would love to have a choice of cheap, fresh fruit throughout the year but she does not.

The truth is that although so much of her time is spent

buying and preparing food she gives little thought to any of it. She scrubs the earth from the carrots and potatoes, cuts the wormholes from the apples and pulls the yellowing leaves from the cabbages and her family eats a reasonable diet. As with the farmer, so it is with her: this is how it has always been. Her only concern is that there should be enough food to keep them all fit and full. The idea that there might be anything wrong with any of it simply does not occur to her. She trusts it because there has never been any reason not to. In years to come she will realise what she has lost.

March 1985
The same farm

The eight-acre field has disappeared. It is now part of a fifty-acre field. The old oak with its car tyre swing has gone and so have the hedges that once marked the boundaries of the field, ripped out by powerful machines with great steel grabs. You can tell the age of a hedge by taking thirty paces, counting the number of different species in that section and applying a formula that seems to differ depending on which expert you speak to. But by any formula this is an old hedge. Some of the trees that formed it were being planted by peasants when Good Queen Bess was despatching Drake to see off the Spanish Armada. Hawthorn and spiky blackthorn, beech and hazel, joined over the centuries by dozens of other invaders, seeded by resting birds or blown in the breeze. Every year the hedges were trimmed and, from time to time, re-laid by men using skills passed from one generation to the next. These hedges did more than mark boundaries and fence in wandering animals; they were part of a vital ecosystem, providing shelter and homes to countless small mammals and birds and insects. Now they are gone, ripped out to the roar of a massive diesel engine and dumped on blazing bonfires. Many of the insects and small animals fled to find a new home. Some perished in the flames. It took two men a few days and it was all over.

An observant visitor might spot the old boundaries because the corn does not flourish along the lines where the hedges once grew. It is as though they had cast their own dying spell on the land where they had stood for so long. The real explanation is more prosaic. The soil along the hedge line is rockier and thinner than the rest of the field. This is where generations of farm labourers dumped stones plucked from the centre of the field. As the years wear on that last reminder will disappear.

The destruction of the hedges to create much bigger fields is only one small part of the changes that have taken place in the past forty years. This is no longer a mixed farm. The dairy herd has gone these many years past. The old milking parlour has been knocked down and most of the cows themselves have been shipped off to the slaughterhouse to end up as stewing steak. Some of the younger animals were bought by the nearest dairy farmer in the next county. The rest were either too old or they simply did not deliver enough milk to justify their existence. The few who survive are expected by their new owner to earn their keep and pour more milk than ever before into the collecting jars in their new, computerised milking parlour. There is a price to be paid for all this milk – and it is being paid by the cows themselves. Their udders are grotesquely distended. Their feet and joints cause them great pain, damaged by the massive amounts of protein they must eat to keep up production. No matter, they will be slaughtered in three years or so and new young heifers will take their place. If the strain on their bodies means they grow ill or develop mastitis more often, again no matter, antibiotics will cure them.

Under this new regime they are eating different food. The crushed, home-grown oats their grandmothers and great-grandmothers were fed would never enable them to produce all that milk. Instead they eat sweet nuts made of concentrated protein, delivered to the farm every month and blown into great hoppers above the milking parlour. Computer chips on collars around their necks tell the mechanism the precise amount of food to be released into their bins. The best milkers get the most feed. The farmer who bought the nuts has only the vaguest idea what is in them. All he knows is that the cows like the stuff and milk well on it. Indeed, he rather likes it himself; he often nibbles on one of them. It will be a few more years yet before he discovers what the millers have been mixing into the feed and the disastrous effect it is having on the animals – but by then it will be too late.

The pigs have gone from the farm too; there were simply too many years when they made no money. So have the chickens. And so has the orchard. The trees were as productive and the apples as tasty as ever, but the market for them had disappeared. They could not compete in price or quantity with foreign apples. At first the farmer had found that hard to believe. How could it possibly make sense for apples to be grown thousands of miles away, shipped at great cost to this country and still be cheaper? But the local shops that had once bought his fruit had closed down, driven out of business by the vast supermarkets. He had tried selling to the supermarkets in his area but they had shown no interest. Indeed, he had been treated with ridicule.

Did he not understand that everything was bought centrally and delivered in bulk? Yes, it cost money to transport apples from New Zealand or South Africa, France or South America, but they could be stored for months at a time in special chambers flooded with carbon dioxide to keep them fresh. Or at least make it appear that they were fresh. No need to tell the customers they were eating old fruit. Anyway, they had said, don't blame us. Blame the shopper. She wants what we offer. Look at the checkout queues if you don't believe it. The farmer finally bowed to the inevitable one sad autumn when his trees produced their usual crop and he calculated that it was cheaper to leave the apples rotting on the ground where they fell than pay to have them picked and sold for a few pence a pound. Later that autumn he hired some heavy machinery and grubbed out the trees.

But it is not only the animals and apples that have gone. So have the men who worked here. The hedges have been ripped out mainly so that the biggest machinery can be used. For the price of one of these vast new combine harvesters you could have bought a small farm only a few years ago, so speed and efficiency are what count. You don't want these behemoths confined to piddling little fields, having to twist and turn their way around and wasting valuable time. After they have done their job of cutting and threshing the corn, they leave the straw behind to be rolled into great, round bales by another machine. When the new corn is growing it will be sprayed by booms as much as twenty-four metres wide from the end of one of their skinny arms to the other. And each machine has just one operator, sitting in air-conditioned comfort high up in the cab.

For most of this country's history the biggest employers of labour by far were the landowners, big and small. That began to change nearly two centuries ago. Towards the end of the last century the battle between man and machine was finally over. Now farms of thousands of acres could be run by no more than a couple of men. In the old days they might have employed hundreds.

When the first primitive threshing machines and winnowing machines arrived on Britain's farms in the first half of the nineteenth century many workers rebelled. Life for most farm labourers was tough enough as it was. The myth of happy farm labourers toiling cheerfully away under a blue sky, leading simple but healthy lives with few of the cares of the modern world, could scarcely have been further from the truth. Poverty in rural England was, if anything, even worse than in the most deprived areas of the great cities. They might have breathed cleaner air, but that was about their only advantage. They lived in dark, damp hovels and most were paid only what they could earn during the spring planting and the autumn harvest. Many feared that if these machines took away that work – miserably paid though it was – they and their families would starve. All too often they were right.

Gangs of men got together to break into the farmers' barns at night and destroy the feared machines with clubs and hammers, or they packed bales of hay beneath them and set fire to them. Some landowners bribed them to leave the machines be. Others brought in the military and the Riot Act was read to desperate men. Most gave in with no more than a curse, but some fought to protect their miserable livelihoods. Fields were stained with the blood of men armed with pitchforks and staves of wood battling against redcoats armed with muskets and bayonets. There could be only one victor. Some of the rebels were deported to Australia. Many ended up on the gallows. And the machines went into action.

By the spring of 1985 the number of men employed in agriculture across England and Wales is lower than it has ever been – scarcely a tenth of the figure forty years earlier, and still falling steadily. Running a small, mixed farm with livestock as well as crops still takes a lot of sweat and muscle but there are fewer and fewer such farms in England. A quarter of a million mixed farms have disappeared and this is only the beginning. On the farm where the two boys swung on their car tyre from the old oak in those postwar years, the change is typical of the national story.

The old man who owned it was taken ill in the late fifties and was told by his doctor that his farming days were over. He was reasonably content with his life's work and happy that his son would be able to continue it. He knew that the farm was as sound and productive as when he had inherited it from his father at the turn of the century. It had not made him rich but he and his family had never wanted for anything. He was as proud of what he had achieved as the next man and confident that his son was inheriting not only a good business but a way of life that he valued. He was not to know that within little more than a decade of his death the farm would have changed beyond recognition. It was to become, in the language of the day, an 'efficient' farm.

At first the son was reasonably content to carry on much as before when his father had been in charge of things. There were some things he wanted to do differently, but nothing fundamental. Yet the world was changing around him and he found himself being swept along. He had no complaints. Apart from anything else, he and his family wanted a better standard of living than their parents. They wanted a decent car, or two, and foreign holidays and even private education for the children. He wanted to make more money to pay for it all. The way things turned out, money was to be no problem.

As he took over the running of the farm two seismic shifts were taking place: the growth of the supermarkets, which changed the face of every high street in the land and the shopping habits of every family; and the domination of the Common Agricultural Policy.

Under the CAP farmers would no longer grow only the food they knew they could sell, as they had done throughout history. From now on it would make no difference whether there was a genuine market beyond the farm gates or not; there was always somebody who would buy what they grew. The new customer was Brussels and the cheque was signed by European Commissioners. The age of subsidy had arrived. Its effect was to be devastating.

Britain's isolation in the war years led to the development of intensive agriculture. It was an understandable response to a national crisis and the fear of what might happen if there were to be another war. The CAP, in a Europe moving to ever closer union, had no such justification. It was dreamed up by politicians anxious to pander to powerful farming lobbies and it was run by bureaucrats who could not – or would not – see the appalling damage it was doing. Instead of farming for need, it led to farming for greed. Once it was the weather and the soil conditions that determined the size of a farmer's bank balance; now it was the political calculations of a group of European ministers and the stroke of a bureaucrat's pen.

Almost no part of rural Britain was left untouched by the CAP. Everywhere farming practices were on the point of changing. The damage to the rural environment would be incalculable and obvious for all to see. The damage to the quality and even the safety of our food would be more difficult to measure and more insidious.

It was the CAP that had encouraged the young farmer to abandon the farming methods that had stood his father in good stead and his father before him, to concentrate on one or two crops and to rip out his hedges. That was the way to make the most money. The more acres of his land that qualified for grain subsidies, the bigger the cheque from Brussels. The more grain he could produce, the fatter the subsidy. And if the grain ended up in a vast shed somewhere, unwanted by any buyer other than Brussels, so be it; it was nothing to do with him. He just grew the stuff. And how it grew. The harvests set one record after another. His father had been ecstatic if his fields of cereals had delivered two tons for every acre he sowed. Now the son expects nearly double that and he gets it.

In part he has the plant breeders to thank for it. Modern varieties of corn grow faster than the old ones. They start to flower when the stalks are still short and the ears begin to form when the grass is only a couple of feet high. If the old farmer had planted the same seed the weeds would have had a wonderful time – growing as tall as the corn and blotting out its light. But now there are no weeds and little danger from other pests: sprayed into oblivion, the lot of them. The first dose of chemicals is sprayed onto the field in the autumn, then another early in the spring, then a selective herbicide to kill off any wild oats or cleavers. After that, a growth regulator to make sure the straw does not keel over because it grows so fast there is no strength in the cell walls.

The reason all these chemical props are needed is because so much man-made nitrogen is plastered on the field. Nitrogen increases the water content of a plant and reduces the thickness and, therefore, the strength of the plant's cell walls. That not only makes the plant fall flat but increases the chance of fungal diseases. It also makes the most competitive weeds grow even faster and more aggressively. The sappier plant is higher in sugar and that makes it more attractive to aphids which, of course, have to be killed. So the farmer is engaged in the most extraordinary spraying spectacular: herbicides, fungicides, insecticides, growth regulators and straw stiffeners. And all because so much artificial nitrogen has been applied to the field.

These fields never get the chance to recover their natural fertility. Before the war farmers rotated the crops, seldom allowing the same crops to grow one year after another because the yields would quickly begin to fall. In 1985 there is an answer: man-made fertiliser. A lot of it. So the tractors – powerful four-wheel-drive beasts – are seldom off the field. When they are not pulling sprayers, they are pulling spinners: small pellets of nitrogen or other chemicals flung into the air in wide arcs to cover every inch of the field – four times in the year and at least a quarter of a ton for every acre of wheat. The reward for all that in a good year is three or four tons of wheat per acre. Never, in any farming system in history, has so much energy been pumped into growing food. It takes up to five tons of oil to produce one ton of fertiliser.

The two small boys who swung on the car tyre are now married and one of them has children of his own. Unlike his mother, his wife shops in the supermarkets. She can't imagine how her mother managed to find the time to shop every day. She does hers in bulk once a week. True, it's a round trip of nearly twenty miles and the traffic is dreadful and she hates doing it, but even if she wanted to shop locally it would not be possible. Most of the local shops have closed. She never has to cut a wormhole from the fruit, nor worry about scabby skins or earth on the carrots. Everything is clean and uniform and most of it comes wrapped in plastic bags. She supposes that she prefers it this way. Her mother complains that the apples don't taste of anything, and why can you get only two or three varieties nowadays? But look at all the other things you can buy. Strawberries in December and mangoes all year around.

There is another big difference between her and her mother. Unlike the old lady, she spends a great deal of time worrying about the food she is feeding her children. If there isn't one scare there's another. And who is she meant to believe when some experts say all those chemicals are bad for you and others say there's nothing to worry about? Although the fruit and vegetables look so clean she goes to a lot of trouble to wash everything carefully. In fact she has recently taken to peeling the apples as well . . . just in case. It's not her and her husband she's worried about so much as the children. All those chemicals. She sometimes wonders if the chemicals are affecting them in some way. They seem to be allergic to so many more things than she was as a child. Still, they're growing big and strong – they'll certainly be taller than either her or her husband – so she supposes she's probably worrying unnecessarily.

Her husband sells animal feed to farmers in the county and he is beginning to worry, too. He occasionally brings home stories of dairy farmers whose cows are behaving strangely, having what seem to be fits and falling over in the farmyard. Many of them have had to be put down. No one seems to be quite sure what to make of it all.

March 2020

The same farm

The old farm has disappeared. Its owner did indeed make a great deal of money, even after the gravy train of the CAP had finally run out of steam. He decided he'd had enough of farming and sold a large part of it to housing developers. There was so much demand for new homes in that part of England and successive governments had relaxed the planning laws to such an extent in the first decade of the new century that he was able to cash in. The rest of the farm was bought by a merchant banker from the city who wanted a big house in the country. What he did not want was a farm, so he kept a few acres around the house and the remainder of the land was split up into separate lots and bought by adjoining farmers. The number of farms in Britain has fallen even more dramatically over the first twenty years of the twenty-first century than they did in the closing years of the twentieth. Many of their owners were victims of the vicious recession that had begun in the late nineties.

Apart from a few hill farms and livestock operations, very few landowners in 2020 do anything that would have been recognised as farming in the middle of the last century. Their fields are ploughed and sprayed and harvested by contractors who criss-cross the country, moving from one large farm to another, like panzer divisions crossing Europe in 1939, subjugating the land briefly and then moving on to a new conquest. The most sophisticated contractors use equipment hundreds of miles up in space: geo-stationary satellites which monitor the condition of the fields and the crops and send instructions to the computer-controlled machinery down below. They can even tell which bits of the field are likely to produce the biggest harvest and instruct the fertiliser spinners accordingly. The tractors and sprayers and combine harvesters are all fitted with sensors so no humans need be involved directly in the operation.

Two centuries ago every blade of grass and ear of corn was harvested by men with scythes and sickles, every sod of earth ploughed and broken up by animals guided by men. Now it is rare for a human ever to set foot on the soil. It is rarer still to hear a bird or see a butterfly. This is a sterile landscape. The one thing that appears unchanged is the soil. To any casual observer driving past it seems much the same as it always has, but the men who worked these fields a century ago would spot the differences immediately and would be appalled.

When they ploughed here the soil would fall away cleanly from the blade. The plough would leave firm furrows standing behind it, unbroken until the next machines came along to turn it into a rich tilth. Now the furrows collapse behind the plough and the old trash cannot be properly buried and there are no birds because there are no worms. The soil is thin and poor. The teeming population of micro-organisms that thrived over the millennia has been virtually wiped out. Now that it is too late the scientists are beginning to discover the link between healthy soil and healthy plants. Yes, the wheat and the barley still grow. But what the soil can no longer provide has to come out of a bag – more of it than ever. The earth is now little more than a growing medium. Chemical fertilisers are plastered onto the land as never before but, worryingly, the yield from the harvests has begun to fall, even when the weather is at its most benign. So, still more chemicals have to be spread, much of which finds its way into the streams or sinks down into the water table that feeds the reservoirs.

The seeds sown on this farm – as on almost every other farm in the country – have been genetically modified. When the first GM crops were developed towards the end of the twentieth century most farmers had been delighted at the potential benefits and angry at the resistance of the protestors who wrecked their trial crops. They believed what the biotech companies promised them. They would be able to spray the most powerful poisons when the plants were young and tender so that the weeds and insects would perish, but the crops would be unharmed because their genes had been modified. They would be immune to the poison. Indeed, they would produce toxins of their own to repel any nasty bugs or fungi that tried to make a meal of them. For a while it worked. But things began to go seriously wrong a few years ago.

Not only have the higher yields promised by the seed salesmen failed to materialise, but the most aggressive weeds and the insects that were so easy to kill in the early days seem to be adapting. No problem, said the chemical companies: just use an even more powerful spray. So they did and, for a while, it worked. But again, the pests increased their resistance and many of them began to mutate. Some simply adapted to the chemicals. Some weeds, contaminated with pollen from the GM plants, developed the same resistance. It was all very worrying. But there are many more things to worry about, not least the quality and the safety of the food being produced.

Only one of the boys who had played in the old eight-acre field seventy years ago is still living and he worries endlessly about the health of his grandchildren. He is not alone. Everywhere people are fearful of what they eat, of what it is doing to them and their children. For seventy years the nation has been pursuing a drive towards growing more and more food for ever cheaper prices. Now the real cost of that policy is becoming apparent. At the turn of the century it was officially estimated that ten per cent of the population suffered from some sort of food-borne illness every year. Now there are many times that number, and the illnesses are more severe. More worrying still, there are signs of new illnesses caused by bacteria that have mutated as the result of genetically modified plants and against which the human body has no resistance.

Hospitals are struggling to cope. Of course going to hospital has always been a risky business because of all the bugs in hospital wards, but now there are many superbugs against which the old antibiotics are useless. As far back as 1998 a committee of the House of Lords produced a report pleading for caution in the use of antibiotics on farms. Otherwise, it said, 'multi-resistant organisms will emerge without parallel progress in the introduction of new antibiotic classes'. That warning now has a prophetic ring to it, but its wisdom has been appreciated too late. Even minor scratches nowadays can cause infections that create serious problems. Weakened immunity is at the forefront of the public health concern.

There has also been a sharp increase in the number of people with Type 2 diabetes, which is linked to obesity. Years ago it was a disease that mostly hit middle-aged people. No longer. The first real warnings were being given as far back as the close of the twentieth century when figures showed that children in Britain were getting fatter at a frightening rate. By the year 2000 roughly half of all fifteen-year-olds were either overweight or actually obese. That was three times as many as twenty years earlier, and it has shown no sign of slowing down. More and more young people now have Type 2 diabetes with all the associated problems: kidney and heart disease, strokes and even amputations.

Medical experts have been warning for years that children must change their diets, but their voices have been drowned out by the advertising messages from the manufacturers of burgers and fizzy drinks. Twenty years ago a Food Commission survey condemned burgers for being fatty, salty and stuffed with chemicals – some of them contained six teaspoons of fat even after they had been grilled. But nothing changed.

Chemical pollution of food is a serious concern. For decades people have been worrying about the 'cocktail effect' of different chemicals in the body and hormone-disrupting chemicals. It is still difficult to prove any positive cause and effect, but a growing number of children suffer from disorders of the nervous system and there has been a sharp increase in cancers linked to pesticides. Some studies show a fall in IQ levels. For years there has been a suspected link between high dosages of nitrates, transferred from the soil into vegetables and animal feeds and resulting in reduced oxygen supply to the body tissues when we eat the food. Now there are no doubts.

Nor are we living as long as we used to. Throughout history longevity in the richer countries had steadily increased. A man born in Britain in 1900 could expect to live for about sixty years. By 2000 that had increased to seventy-eight and the assumption was that it would keep rising. Not any longer. Again, diet is suspected.

When food experts talk about a 'healthy diet' in 2020 they no longer mean eating five portions of fruit and vegetables a day, as they once did. Indeed, they warn against many kinds of fruit unless we know exactly how they have been grown. It's the same with vegetables. In the 1990s we were advised to cut the top inch off carrots in case they were contaminated with chemicals. Now they warn that all vegetables and all fruit should be peeled – not just washed thoroughly. Even so, they say, we cannot be sure what harm they may be doing us.

The fertility of British men is lower than it has ever been. Sperm counts consistently fall below the World Health Organisation level that says a fertile man should have twenty million sperm per millilitre in order to reproduce effectively.

Many processed and ready-prepared meals which contain additives and preservatives are suspect. More and more people are suing food manufacturers for health problems associated with their products. Some are succeeding. The search for 'safe' food at a price people can afford is now becoming a national obsession. Supermarkets, terrified at the effect of all this on sales and profits, are beginning to assume many of the functions of an over-stretched NHS. Just as Iceland banned GM products in 2000, they are beginning to ban various foods containing certain additives or grown with too many chemicals. All schools include lessons on the dangers of eating the wrong kinds of food.

The government is being urged to order manufacturers to put warning labels on packaging, just as the tobacco manufacturers were ordered so many years ago. The message is terrifyingly simple: Warning! Eating may be dangerous for your health.

So is that really where we shall be in a generation from now? Probably not. I have taken all the warnings, all the worries, and added them together to paint a nightmare picture. There are, of course, many other pictures I could have painted based on how the available evidence is interpreted. The biotech companies insist that genetic modification will, indeed, deliver the goods. They say some modified crops are already achieving good yields with less fertiliser and fewer insecticides and herbicides, so the risks from chemicals will be smaller rather than greater. Perhaps.

More farmers are converting to organic farming methods and using virtually no chemicals and others are changing their farming practices to use fewer. Some of the nastiest chemicals – lindane, for instance – will be illegal by 2002. Livestock farmers are being forced to cut down on antibiotic use.

So it could be that farming and food production in Britain will look very different indeed a generation from now and our food will be safer and more nutritious than it has ever been and we shall wonder what all the fuss was about. I hope so, but I doubt that too. The economic pressures on farmers for bigger yields is still great. Many of the biggest arable farmers refuse to accept that they are doing any harm to the soil by treating it as nothing more than a growing medium. It has produced bumper crops for half a century, they say, so why should it stop now?

The agrochemical companies whose profits and share price depend on selling more and more of their products are still a powerful force. And, most disturbing of all, there is still reluctance by politicians to accept that we are running any real risks. In the chapters that follow, I shall examine those risks in greater detail. But first, let us go back in time and look at our relationship with the food we eat.

FROM CAVEMAN TO KITCHEN

The History of Food

Early humans could have taught us a thing or two about healthy eating. Their diet had plenty of good stuff in it – fruit, berries, green leaves, roots, nuts, seeds – and it was all eaten raw. So there was plenty of fibre, vitamins and minerals, carbohydrates and nutrients and it did us proud for millions of years. There was just one problem. Too often there was not enough of it. So when the food ran out, we had to move to where there was more or starve. An empty stomach was a powerful incentive for colonising new territory.

Things picked up when our ancestors discovered, a few hundred thousand years ago, what happened if you rubbed two sticks together for long enough. Fire was one of the big turning points in the evolution of the human diet. It not only made mammoth steaks that much tastier and enabled people to eat things they could not have eaten without cooking, it also made it possible, many thousands of years later, to preserve meat by smoking it.

When they got fed up with having to chase and catch their dinner, they began to domesticate it. Herding techniques were developed 20,000 years ago so that animals could be reared and bred in captivity for their milk, meat and skins. And then, 10,000 years later, came the biggest breakthrough of them all – the realisation that it was possible to grow food in our own back yards rather than have to trudge off to the forest to pick whatever happened to be available. This was man controlling nature for the first time, selecting what was to be grown and where. Agriculture had arrived. It took another few thousand years for the next really big change in agriculture – the cultivation of grain – and that was about it. Our diet was pretty much established. We may have added a few bits and pieces over the years, but put aside such civilising developments as pot noodles or pizza and the absolute basics have not really changed all that much for a few millennia.

Something else that was established 20,000 years ago was the connection between food and health. When we started seriously interfering with nature by rearing animals in captivity we began to get sick in new ways. Hunting mammoths with wooden spears was a risky business but so, it turned out, was living with rather less ferocious animals. We started to catch their diseases, to become infected with their viruses and their bacteria. Pigs and ducks gave us flu; horses gave us colds; cows gave us the pox and dogs gave us measles. Even so, we British loved our meat and we ate vast amounts of it – as much as we could get hold of.

For almost all of our history the single greatest preoccupation of man had been to find, kill or grow enough food for himself and his family. It was the force that drove him to conquer new lands, to burn down forests, to tame wild animals. Once there was enough food for most people most of the time the emphasis began to shift. By the seventeenth century we began thinking about what effect food was having on our health: not just the occasional herb, but our diet as a whole. This was a new notion. In the fifth century BC Hippocrates had stated: 'Let food be your medicine and medicine your food.' In ancient texts plants were listed for their curative properties. The value of herbs had been known for millennia, and the idea that an apple a day kept the doctor away was popular long before anyone had ever heard of antioxidants. But the notion that food itself was more than something to fill our stomachs, that the right mix of nutrients, minerals and vitamins was vital for a long and healthy life, took much longer to achieve any kind of scientific credibility. It was the sailors of the British navy who provided the first clues.

Offhand it's hard to imagine a more hellish life than sailing the seas a few centuries ago. It seems a minor miracle that any sailors ever survived. For months on end they lived in disgustingly insanitary conditions, crammed below decks in stifling heat, often punished savagely for relatively minor infringements, and subsisting on a diet of salted meat, biscuits and rum. The only thing that was fresh were the weevils that infested the biscuits. No wonder so many expeditions to the further corners of the globe failed. Sometimes the ships came to grief in storms or on uncharted rocks but far more often it was because so many men fell ill and died. They feared scurvy more than any other illness. It killed more men than all the storms and battles and floggings and hangings combined. It was a foul disease that weakened a man until he could no longer stand, his body covered in pustules, his breath smelling of decay and death. And it took a Scottish physician – James Lind – to conquer it.

It is not only the eighteenth-century sailor who owed his life to the brilliant doctor; we are all in his debt. Lind did some work in 1747 on board the Royal Navy vessel *Salisbury* that was probably the first example of its kind of a controlled clinical study. It was a precursor not only of today's randomised trials but of modern evidence-based medicine. And it was the first time that a nutritional disease – scurvy – had been scientifically investigated. In the treatise that Lind published five years later he wrote a sentence so prescient, so wise, that it should be printed out in letters six feet high today, mounted on a plaque and hung on the office wall of every doctor, every NHS manager and every politician in the country:

'Almost all diseases are easier prevented than afterwards removed.'

Lind had studied the sailors' diet and proved the connection, long suspected, between nutrition and scurvy. He noted the importance in a diet of 'broths made of fresh flesh meats, together with plenty of recent vegetables, if they can be procured' but it was his cure for scurvy that proved to be a lifesaver. He recommended 'two oranges and one lemon every day for six days when visible effects were perceived'. That was it. Scurvy was caused – not that Lind had ever heard of it at the time – by a lack of Vitamin C. Citrus fruits contain Vitamin C. Disease cured. Better than cured: it could be prevented for ever. Sadly for thousands of sailors, it was to be another eighty years or so before the naval diet was changed so that every sailor was provided with a daily ration of lime juice. In the meantime, though, ships began to carry a 'medical comfort' devised by Lind: 'Let the squeezed juice of the fruits (oranges and lemons) be well cleaned from the pulp [. . .] and boiled for several hours, until the juice is found to be of the consistence of oil when warm or of syrup when cold. It is then to be corked up in a bottle for use.'

The remarkable Dr Lind had another claim to fame. He developed 'portable' soup. You might like to try it some time. All you need are the offals, shins and feet of a cow. You boil them for a very, very long time, strain the broth and pour it into moulds where it sets into rock-hard cakes of glue. Then you hack off a bit when you're feeling peckish, boil it up again and . . . disgusting. Or so the sailors thought. But it was extremely nutritious and, until we learned the techniques that made canning possible, not a bad substitute for oxtail soup.

When Britain was a largely agricultural society and most of us lived in small villages there was little danger that our food would be adulterated. Before the Middle Ages the housewife was her own baker, butcher and brewer. So the diet might have been monotonous, but it was reasonably wholesome and safe. Serious adulteration started to happen with the onset of the industrial revolution when we moved from village to town. Once there was a large urban population there was an opportunity for unprincipled manufacturers and traders to move in and make money. Not that adulteration of food was entirely new. The earliest mention of it dates back to Roman times. So many people complained of the wine tasting a bit ropy in ancient Rome that inspectors were appointed to look into it. They found 'artificial maturing of wine with the use of aloes and other drugs'. Roman bakers were accused of adding 'white earth' – carbonate or magnesia – to bread.

The cheats were at it in this country too a millennium later. After the Norman conquest, complaints were recorded about merchants and shopkeepers cheating their customers with underweight loaves of bread by inserting little chunks of iron in them. Worse still, they watered the beer and then added sugar to disguise the taste. It must have tasted similar to what passes for beer in some of the trendier bars of London in twenty-first-century Britain. Probably better. But worse things were to happen to beer and to many other foods a few centuries later.

In 1820 Frederick Accum published a treatise on the adulteration of food and culinary poisons: 'There is death in the pot.' This was the first time the subject had been brought out into the open and discussed in a scientific manner. Accum proved that many of the basic foods of the day were adulterated: bread, beer, tea, wine, confectionery and condiments. Bread was one of the first. London bakers used alum, a mineral salt which made cheaper bread look whiter and, therefore, more desirable. The most expensive bread was the whitest because it came from 'first flour'. The alum had some pretty unpleasant after-effects. Although it was not poisonous it made digestion difficult so the nutritional value of the bread was lowered. No less a figure than Tennyson commented on it: '. . . chalk and alum and plaster are sold to the poor as bread'. But what the nineteenth-century hucksters did to drink was even more dangerous.

In 1819 more than a hundred brewers and brewers' druggists had been convicted of using various substitutes for expensive malt and hops in porter and ale so that it could be diluted but still taste strong and get you drunk. It did more than that. One of the substitutes was a nasty poison called cocculus indicus. In the years that followed dreadful things went into the nation's food and drink and the people responsible were seldom called to account. Andrew Wynter wrote an account of a breakfast that he and a party of friends had 'enjoyed'. The food included 'highly flavoured putrid meat, the coffee contained dried horse's blood supplied by the knacker's yard and the cream was thickened by calves' brains'.

The adulteration was happening across a wide range of different food and drink: water in milk and beer, chicory in coffee, barley meal in oatmeal, copper and lead carbonate in sugar. And the scale of it was so great that it was seriously affecting the health of city people, especially the poorest. Children reared on a diet of adulterated bread and diluted milk were less able to fight off the many infectious diseases and gastric complaints that threatened their frail little bodies. Infant mortality was appallingly high. Some children were poisoned by mineral dyes in sweets; adults were paralysed by lead in cayenne pepper and snuff. And the effects were cumulative. Trace elements of lead, copper, mercury and arsenic built up in the body over time and it is thought they may have caused chronic gastritis, the most common disease of the urban population in the early nineteenth century.

If many people were getting ill or dying because of what

was happening, those responsible were getting rich off the back of them. When they were eventually called to account some traders put forward a justification that may sound familiar to us a century and a half later. In effect, they said: 'We're not doing it to line our own pockets. We're doing it because it is what the public has asked us to do. It lowers the price of food and that's the only way the poorest can afford to buy the things they want.'

Until about 1850 that disgraceful justification was more or less accepted by most people. Those who could and should have known better also managed to persuade themselves that adulteration was not really all that widespread – just confined to a few districts of the bigger cities – and, anyway, it probably wasn't really harmful. Then Thomas Wakeley arrived on the scene. He was a doctor and a radical, the editor of the *Lancet* and the coroner for West Middlesex, and he had seen too many patients who had died or suffered at the hands of the food adulterers. He demanded that there should be a thorough investigation and Dr Arthur Hassall, a physician and lecturer at the Royal Free Hospital in London, was appointed to conduct an inquiry.

Never before had such a rigorous series of tests been applied to so many different kinds of food. A total of 2,400 analyses had been carried out by the time the inquiry ended. The findings proved beyond doubt that if there was money to be made out of adulterating a particular food, then that is what would happen. By then it was almost impossible to buy any basic foodstuff in its pure state. For once, the identity of the guilty men was not protected. The names of the manufacturers and the traders were published for all to read. This is how an article in the *Quarterly Review* described the findings: 'A gun fired into a rookery could not cause a greater commotion than did the publication of the names of dishonest tradesmen; nor does the daylight, when you lift a stone, startle ugly and loathsome things more quickly than the pencil of light, streaming through a quarter-inch lens, surprised in their native ugliness the thousand and one illegal substances which enter more or less into every description of food which it will pay to adulterate.'

Punch joined in the campaign with its own inimitable style. A series of savagely satirical articles began with a midnight party: 'Imps of all trades were there. The baker imp who grinds his alum . . . the grocer who enriches his chocolate with brick dust . . .' Every time Hassall published another exposé, *Punch* responded with another powerful parody. Parliament began to take notice. Questions were asked about why there was no adequate food legislation and why people were not being protected. Parliamentary committees were established. Throughout 1855 and 1856 MPs heard evidence from doctors, chemists, traders and manufacturers. Dr Alphonse Normandy, the author of *A Standard Work On Commercial Products*, summed it up before the committee in this way: 'Adulteration is a widespread evil which has invaded every branch of commerce. Everything which can be mixed or adulterated or debased in any way is debased.'

The final report of the committee concluded: 'Not only is the public health thus exposed to danger and pecuniary fraud committed on the whole community, but the public morality is tainted and the high commercial character of the country seriously lowered both at home and in the eyes of foreign countries.' It could only be a matter of time before new laws were brought in.

One bill after another was introduced in Parliament and, year after year, MPs and peers with financial interests to protect managed to get them thrown out. The turning point came in 1860 when the Adulteration of Food and Drugs Bill became law. For the first time in history people had a legal right to expect that the food they were buying had not been diluted or adulterated. Over the years that followed there were to be more new laws protecting people against unscrupulous traders and manufacturers. No longer could food suppliers claim that they were adulterating their food because that was what the public wanted. For the first time people were legally protected against 'death in the pot'. But the battle over precisely what goes into the pot and onto our plates continues to this day.

A century after that first Act on food adulteration became law the publication *Dear Housewife* carried an article from one Doris Grant, an advocate of the 'natural school' of food. 'Do you know,' she wrote in 1959, 'that there is hardly an honest food left to buy? Nearly all the foods that go on to our table are so changed, so processed and chemicalised, that all their original goodness is either removed or killed. They are bleached, dyed, dehydrated, frozen, synthetic, tinned, sulphured, pasteurised, iodised, refined, adulterated and too often unclean as well. It seems a wonder that we are alive at all.'

Mrs Grant might as well have saved her ink. As the twentieth century neared its close it was even more difficult to buy her 'honest food'. Professor Erik Millstone, who heads the science and technology research unit at Sussex University, is one of Britain's leading authorities on what happens to our food before it reaches the supermarket. He estimated that no fewer than 3,850 additives are in use and he calculated that the average person eats an astonishing four kilos of additives every year. On that basis a baby girl will have eaten her weight in additives by the time she is seventeen years old.

Professor Millstone believes that of all the major industrialised countries Britain has the weakest set of regulations governing food additives. There are, he says, quite a few that we permit although they are banned in other countries. The justification for such a long list is that a variety of chemicals can be used to accomplish each individual task and, as a result, less of each is used. Millstone is not impressed. He regards that as 'a device of imaginative rhetoric that deserves a prize'.

Perhaps it is not so astonishing if you take a look at the tins and packets on your kitchen shelves today and read the labels. You will not find too much arsenic listed, nor alum in the bread, but nor is it likely that you will understand the significance of all the different chemicals and additives listed. We are frequently assured that the chemicals have been tested and are not harmful to our health. Yet some toxicologists believe that not all have been properly assessed and in too many cases the industry is given the benefit of the doubt. There is also the question of how we should define 'harmful'.

Many of the additives will be used because they change the taste of the food. The manufacturers know how to appeal to our taste buds and even how to manipulate them. We become conditioned to expecting certain tastes – just as if you feed a baby sweet foods he will resist more savoury fare. Vast amounts of sugar will be added to appeal to the sweet tooth, never mind that a frighteningly large number of children are now much too fat. Obesity is one of the biggest killers in an affluent society such as ours.

One of the main causes of cardiovascular disease is too much salt in our diet. Take another look at those labels and see how much salt has been added. True, the supermarkets have grandly declared that they have instructed their suppliers to use less salt in some of their products. Food manufacturers advertise their 'healthy' beans, or whatever the food may be. Why don't they make all their products 'healthy'? Because they fear they would lose sales to their competitors. It is a purely commercial calculation and nothing to do with health at all. Some of the additives are needed if we want food to last longer than nature intended, as indeed we do. There is even a legal requirement to add some chemicals. It's against the law to sell dried apricots unless they contain sulphur dioxide, for instance. And not all 'E numbers' are bad: E300 is ascorbic acid, an important antioxidant. Others are less desirable. It is almost thirty years since an America doctor, Ben Feingold, warned that the accumulation of food additives provided the main trigger for hyperactive behaviour in children.

One way or another, then, our food has been messed about with over the millennia for reasons both good and bad. Meat that has been smoked or salted may not be as good for us as fresh food but if it's a choice between that and starvation there's no contest. Our ancestors had no choice in the matter. Even if they had known about carcinogens they'd have still smoked and charred their mammoth steaks. Food was survival.

As recently as half a century ago, when I was growing up as a small boy in a working-class household, the same basic principle applied: food was fuel. There were five children and no money to spare and no nonsense about 'enticing' us to eat one thing or another. If we didn't eat what was put in front of us then – like every other child I knew – we went hungry. So we ate and we thrived. Nothing was wasted. Leftover potatoes and greens – if there were any – made bubble and squeak; stale bread made bread pudding. I can remember to this day what we had for pretty well every meal every day of the week. It varied a little according to the season but not much. Modern nutritionists would not have approved.

The meat was usually the cheapest cut – 'scrag' end of lamb to make stew and breast of lamb or brisket of beef to be roasted – and consequently it was very fatty. Chicken was a luxury: once or twice a year on special occasions. The idea of eating five portions of fruit or veg a day, as the government now exhorts us to do, was unimaginable. Even if it was available in those postwar years it was simply too expensive. We ate apples in season (often scrumped from the gardens of middle-class homes) and blackberries from dusty hedgerows and, very occasionally, oranges. That was it. Fruit juice did not exist except as a form of orange concentrate served in school, mixed with water. Vitamin supplements came from cod liver oil, which was disgusting, and thick, almost black, syrupy malt, which was delicious.

Today choice seems to be all that matters. A large supermarket will stock around 15,000 food items and recognises no seasons. That link – and the link with the farm that produces the food – has long since been broken. But food is relatively cheap. The proportion of a modern household income that goes on the weekly food shop is a fraction of what it used to be. All but the poorest families have a more varied diet than we have ever eaten. Yet something has gone wrong.

The British population was healthier in the 1940s than it had ever been before. Ridiculous as it now seems, when the Attlee government created the National Health Service the budgets were calculated on the assumption that costs of health care would actually fall. It is true that we live longer than we have ever lived and life expectancy continues to increase. But the bald statistics are misleading. After we have reached the age of fifty life expectancy is much as it was a century ago. In those days if we managed to survive a dangerous childhood we would ultimately be knocked off by something like pneumonia or bronchitis or other respiratory diseases. Now older people suffer from a wider range of unpleasant illnesses. There is more cancer, diabetes, heart disease, degenerative illness than there used to be. Our diet is crucial to all this. Good food is vitally important. But what is good food? It seems to me there are four conditions that need to be met.

- Food must be safe. That's hardly a scientific word but I use it to mean that it should be as free as possible of bacteria that might poison us and chemical residues that might build up in our bodies and create problems later in life.

- It must be nutritious. A balanced diet should contain all those essential, primary nutrients, vitamins, minerals and trace elements that build a healthy body.
- It should also contain the secondary nutrients about which we know less: for instance, the flavenoids and metabolites present in healthy plants with a strong defence system of their own. We now know that they possess antioxidant properties that help us fight cancer.
- And finally, healthy food can come only from healthy plants and animals. We have seen what happens when we treat cows or chickens with contempt, as no more than food machines in food factories. They get ill and we get ill as a result. I make no apology if this sounds terribly unscientific and subjective, but I believe there is a life force in any healthy living creature or plant. Some call it an 'organising principle'. You cannot measure it but you can see it in a healthy, happy baby or in an oak tree in its prime. You can sense the energy and you want to tap into it. You feel more alive just watching a small baby, arms and legs going like pistons, delighting in life. Really good food should burst with taste and freshness and vitality. You know it when you eat it and you feel better for it.

Modern, industrial agriculture is failing us in most of those respects. Many of us have become suspicious of food – no matter how brilliantly it is marketed and how beautifully it is packaged – and with good reason. That is sad, but if it means we gain a new respect for food it is also to be welcomed. Eating is one of the most intimate of activities. We learned to kiss – or so we are told – when mothers chewed food and passed it to their babies with their mouths. We must be able to trust our food, for ourselves and our children, and treat it with respect at every stage, from the field to the plate. When that trust goes we lose something precious. Science has proved what our intuition has always told us: our health and our wellbeing are intimately bound up with our diet.

As a scruffy postwar youngster with a heroic appetite none of those thoughts occurred to me. Food was food and the more of it there was the better. My mother bought what she could afford, confident that it was nutritious and safe. She trusted it. I doubt she would be as sanguine if she were still alive and feeding a young family today. I know I'm not. Nor are most of my friends. Even those who scoff at some of the worries I have been voicing make a point of buying food for their babies that has been grown without man-made chemicals. Within the next year or so all the baby food sold by the big manufacturers will be organic. A generation ago they sold none. That alone suggests a level of unease that should give everyone pause for thought.

In the rest of this book I shall be exploring the reasons for that unease. Is it justified or are we over-reacting as a result of some pretty nasty experiences? In the next few chapters I want to look at the first of the conditions for good food that I outlined above. Can we assume it is safe and if not, should we blame the farmer?

THESE TOXIC TIMES

Pesticides

The problem with writing about farmers is that they can be such sensitive souls. Their reputation for always complaining about something or other has not come about entirely by accident. The weather is either too wet or too dry, too hot or too cold. The politicians either interfere too much and don't let them get on with their job, or they ignore their difficulties when they are pleading for the government to help. When times are genuinely hard farmers complain that bankruptcy is just around the corner and when they are making lots of money they say they could make far more if they flogged the farm, put their feet up, and lived off the interest from the building society. And they get even more cross when they hear city people endlessly parroting the refrain: 'When did you last see a poor farmer?'

The truth is, of course, that there are an awful lot of poor farmers and some of them are very poor indeed. Those, it goes without saying, are not the great barley barons with their thousands of acres of flat, rich, black soil and their bank balances as fat as a grain silo in September. They are the hill farmers of Wales or Cumbria with land so steep only a mountain sheep feels really at home and soil so thin it may grow a few blades of grass if the weather is really kind but that's about it.

They are the small dairy farmers who borrowed up to the hilt to buy a bit of extra land or a fancier milking parlour when milk was delivering serious profits and now can scarcely afford to pay the interest, let alone meet all their other costs.

They are the tenant farmers who have never owned their land and have to pay rents out of income that has shrunk to almost nothing.

There are many of them and the only reason some survive is that the banks, who have forced them off their land in past farming crises, have realised that if they do the same again they will cause a crash in land prices and hurt themselves in the process. Many of the smallest farmers who manage to survive live lives of such poverty that if they were on some benighted council estate in a big city the papers would be full of their hardship. Perhaps country poverty seems somehow more genteel, less brutal. It is not. The farmer may look out on fields and hills instead of litter-strewn streets and abandoned car wrecks but real poverty can blind your eyes to beauty. As one old man, who had seen plenty of hardship in his time, put it to me when I stood admiring the views from his front porch: 'Aye, but you can't eat a bloody view, can you?'

When things are going really badly for a farmer there seems no escape. Everywhere he looks he sees failure. He sees the lambs or the beef cattle that he knows will bring him less money at market than it has cost him to rear them. Or he sees the tractor that is so old and has been repaired so often it cannot possibly survive another season. Or the crumbling walls or collapsing fences or broken gates hanging off their hinges. Or the water-logged field that desperately needs draining, the encroaching gorse or bracken that must be tamed. It all costs money or needs labour. He has no money and he is running out of energy.

He sees reproach in the eyes of his children who want a holiday or new trainers that he cannot possibly afford and, even worse, he sees despair in the eyes of his wife who can see no way out of this miserable existence. He could, possibly, sell the farm but the buyers are not exactly lining up and by the time he had settled his mounting debts there would be precious little left. And then what would he do? The only thing he knows is farming and there is no work to be had. And where would he live? In the nearby town, in some dingy rented accommodation, on the dole. The farm has been in the family for generations. His father handed it on to him, just as he used to hope that one day he would hand it on to his own son. When he looks in the mirror as he is shaving he sees a failure. It is small wonder that so many farmers have put an end to it all with a shotgun.

Yes, you can say it's their own fault. They have had better times over the years and many have enjoyed a pretty good life, often at the expense of taxpayers forking out for the subsidies they have enjoyed. And nobody forced them to farm anyway. There is something in that, but anyone who thinks small farmers, as opposed to the big landowners, have ever had an easy life should think again. I've had some experience of my own at the sharp end and I can tell you that nobody works harder than a man running his own small farm.

If it is a dairy farm then, as everyone knows, you need to be up very early indeed for the milking. No problem, you might say: go to bed early. That's what I do when I am presenting the *Today* programme the following morning. But it's not quite the same. When I set the alarm clock in my London home I know I won't have to struggle out of bed three hours later to help a cow give birth to a calf that refuses to be born in the normal way. And on a farm there is always something else waiting to be done. In the spring and summer months there are the fields to be worked and the harvest to be made and the new grass or barley to be sown. The cows must be brought in to be milked from wherever they have been grazing and taken back again afterwards. Twice a day, every day.

Electric fences must be strung up to stop the cows stealing the grass from the grazing that you're saving for tomorrow or next week. If you leave them to graze where they choose they will trample and spread dung everywhere and waste at least as much fresh, clean grass as they eat. That's one of the things about cows: they are never satisfied with what they've got; they always want more. As my mother used to put it, their eyes are bigger than their bellies. When the electric fence is trampled down, as it frequently is because the smarter cows have figured out that the battery has gone flat and the current is no longer flowing, it has to be fixed and the cows chased back again.

In the winter months the cows are kept inside in sheds, so that makes life easier: no time need be spent rounding them up and bringing them in. Ah, but they have to be fed and that means the silage must be cut and hauled. And they make an almighty mess in the sheds and in the yards and that has to be cleared up. Again, twice a day every day tons of cow dung must be scraped and pushed into the slurry pit. And all the things that should have been done in the summer and were left for the winter months must now be attended to. There are fences to mend and hedges to be trimmed and gates to be hung. There are ditches to be cleared and fields to be drained and machinery to be fixed and buildings to be repaired. You never finish work on a farm; the best you can hope for is that you can complete one job before you have to move on to another. Because there is a relentless rhythm to dairy farming that drives and dominates everything else. Twice a day, every day, come hell or high water, the cows must be milked.

Of course there are good times in farming, even if you are running a small farm single-handed as I had to do when my farm manager was away on holiday. At the end of a successful harvest it is immensely satisfying to see the barns stacked high with bales of fresh hay or straw and the silage clamps packed tight with the slowly fermenting grass that will feed the cows through the winter months. Leaving the farmhouse on a spring morning, when the mist is still clinging to the fields and you can smell the dew on the grass, can make you feel glad to be alive. And watching a perfect little calf struggling to its spindly legs and suckling its mother within minutes of being born is always a moving experience.

But those are the sweet moments. Mostly it is just endless hard work – physically and mentally demanding – with precious little reward. For me, it was easy. I had another income to fall back on. If the farm failed to produce a penny profit after all the costs had been met – as indeed it did – it was not a disaster. I always knew that if I chose to sell up, I could do so and return to my other life in London with little more than a backward glance. Not many farmers in the real world of debt and overdraft have that option. If the farm is not profitable, they have no income. If the pretty little calf fetches no more than a pittance when he is sold at the market and the milk cheque barely covers the cost of producing it, then what has it all been for?

It is easy to feel sympathy for the struggling small farmer in the hills and valleys of the west. It is tempting to dismiss the complaints of the barley barons with their thousands of flat acres and giant new combine harvesters. Yet they too must make a profit. The cost of the best land has always been relatively high and they have a huge investment in their property. They may rent a large part of it or owe money to the banks. Farming, when all is said and done, is a business. The debts must be serviced and the wages paid. We should recognise the business reality when we accuse farmers of vandalising great swathes of the countryside and spreading pesticides over every square inch of fertile land. It does not make it right – vandalism is vandalism – but there are two sides to most stories.

Yes, some farmers have been disgracefully greedy over the years, tearing out ancient hedges or copses, draining every last water meadow, ploughing up old pastures or precious chalklands. Some have treated the soil as nothing more than a growing medium, with no thought to the future, saturating it and everything that grows in it with chemicals. But we need to look a little further than even the greediest of farmers if we are to identify the real villains of the piece.

For ten thousand years, since the dawn of agriculture, farmers have searched for ways to encourage their land to produce more food. They soon came to understand that if you take something out by harvesting a crop, you must put something back in by feeding the soil. The exception to that rule is what are called nitrogen-fixing plants, such as clover and legumes, which take nitrogen from the atmosphere and send it down to their roots. But all plants must feed if they are to grow and some are hungrier than others.

Until the last century the soil was the master and the farmer its servant. If it was treated harshly, if too much was demanded of it, then it would simply refuse to yield a decent harvest. The farmer would be forced to concede. Either he would have to feed the soil with enough organic fertiliser to replace what he had extracted – animal or human waste, vegetable matter, seaweed – or he would have to give it a good, long rest on a regular basis. In the earliest days of farming, when most of Britain was still covered in forest, the trees would be burned down to clear the ground. Their ash and leaves enriched the soil. Later the practice was to allow fields or strips of land to lie fallow every third or fourth year. Or the crops would be rotated: plants hungry for nitrogen would follow a crop of clover, which would be eaten by the cows. Over the years, farming practices varied but the basic principle remained unchanged: the earth provided the harvest and, one way or another, the harvest returned to the earth. For the next generation to survive and the one after that, and the one after that, farming had to be sustainable. To take from the soil without giving back and without protecting its richness would be the ultimate betrayal of future generations.

There was another sound reason for rotating crops. As every good gardener knows, if you grow the same vegetables in the same ground, year in and year out, the pests that breed in the soil will love you for it. As with very small children, pests like to know where they stand. Give them a nice, regular routine and they will thrive. So if you plant carrots or cabbages in the same ground one year after another you can be pretty sure that the cabbages will get club root and the carrots will get root fly. It makes no difference whether you are sowing two rows in your back garden or a thousand acres of open farmland, the same rule applies.

Sadly, the same tactics do not work with pests that attack from above. You cannot fool greenfly and blackfly, caterpillars and slugs, by rotating. One way or the other, if they're looking for food they will find it. They are always looking for food and farmers have always been looking for ways to kill them. Insects have been on this earth for far longer than humans or other animals. They are infinitely adaptable and extraordinary survivors. It is easy to find ways to kill some of them but not so easy to make sure they do not return for another bout. Since the Sumerians first discovered that they could kill mites with sulphur in 2500 BC, farmers have experimented with poisons, often killing many beneficial insects, such as bees, in the process and frequently killing themselves. Cyanide and arsenic were particularly popular and arsenic was the first substance to be recognised as a carcinogen. It was linked to cancer more than two centuries ago by an English doctor. Before the last war, cancer specialists in the United States were appalled at the way pesticides containing arsenic were dusted, with careless abandon, across vast areas of the cotton-growing south. And without effect – the bugs lived on to plague future generations of farmers.

It's the same story with weeds. In 1825 farmers were encouraged to pour common cooking salts on the crowns of weeds. It might or might not have killed a few weeds; it would certainly have killed the fertility of the ground for a good long spell. Through it all, the weeds and the pests survived. Some minor victories were chalked up in the long campaign and occasionally farmers might have won the odd battle, but not the war. Instead, the wise farmers recognised that, as with poverty, weeds and insects would always be with us. So the best thing to do – the only thing to do – was to learn to live with them. And that they did.

They knew that they would lose a certain proportion of their crops to pests and the aim was to keep that proportion to the minimum. As the centuries went by, the wisest farmers recognised that nature itself could be recruited in this war of attrition. If hedges were planted around fields, not only would they provide shelter and fencing for the animals, but they would also offer a home to other insects and birds who would feed on the pests. One ladybird can polish off a plateful of greenfly for her dinner. Song thrushes like nothing better than a breakfast of slugs as they sally forth from their night-time shelter to attack the cabbages or young barley. The thrush has the additional benefit of lifting the hard-working farmer's heart with his elegant song, but it's hard to put a price on that.

So, for ten millennia or so, farmers acknowledged – albeit reluctantly – that they were never going to kill off every pest and they might as well learn to live with them. In the last century everything changed and, once again, we had a war to thank for one of the more significant turning points. This time, though, it was the First World War.

*

Throughout the ages military commanders have dreamed of a weapon that would kill the enemy without the arrow or the musket round or the shrapnel scoring a direct hit. Chemical warfare was inevitable.

My own uncle was one of its earlier victims in the trenches of Northern France. He survived, but his lungs were horribly damaged. He was never again able to lie down to sleep, for fear that his lungs would fill with fluid. He died a painful, choking death when he was still a relatively young man. Many of the pesticides used routinely on our farms and gardens for decades past are direct descendants of the nerve gases that destroyed the life of my uncle and millions more like him before they were finally declared illegal in warfare. They have been an extraordinarily potent weapon in the war against pests, attacking their nervous systems just as they attacked humans on the battlefield. Whether the war against the pests has been won, though, is a question that has yet to be answered. And even if it has, it may prove to have been a pyrrhic victory.

DDT – which was to become one of the world's most notorious pesticides – was discovered shortly before the outbreak of the Second World War. It had first been synthesised in the previous century, but it was not until 1939 that a scientist in Switzerland, Paul Müller, discovered how to use it to kill insects. From then on DDT was used for just about everything: from killing many of the disease-carrying insects that had plagued mankind for millennia, to treating soldiers covered in lice, to wiping out pests on the farm. Dr Müller became a hero almost overnight and was awarded the Nobel Prize in 1948.

In Britain the age of pesticide farming began as the Second World War came to an end. Until then, as Oliver Rackham wrote in his *History of the Countryside*, much of England would have been instantly recognisable to Sir Thomas More and even the Emperor Claudius. The urban landscape had, obviously, changed beyond recognition. New towns and villages had been developed and the cities had spread their boundaries, encroaching onto what was once agricultural land. But in large areas of Britain the hedges and downland, the fens and the ancient woodland had remained unchanged over the centuries. Then after the war, for reasons I described earlier, the politicians decided they would take control of agriculture. They wanted to see more food grown in Britain so that we would be less dependent on imports in case of another war. From that point on everything began to change. The strategy employed was to introduce a system of subsidies which put more money in the farmers' pockets. They were persuaded to grow as much food as they could force from the land. The new weapon was to be a vast arsenal of chemicals.

If the name 'Graham Harvey' means anything to most people in Britain it's because he is the man who spent years as the agricultural adviser to the nation's favourite everyday story of country folk, *The Archers*. What he ought to be famous for is his book *The Killing of the Countryside*, a fierce and well-researched attack on the system of subsidies which has done so much to damage agriculture and the rural environment in this country. In the book Harvey described how a new chemical weapon had been secretly tested during the war against Hitler. The weapon was not to be used against Hitler's panzer divisions, but against an army of weeds growing in the fields of Britain. Its name was 4-chloro-2-methylphenoxyacetic acid, mercifully known by the acronym MCPA.

The team involved in the project felt they were taking part in 'a great agricultural adventure'. It was taxpayers' money that paid for it, but the chief agricultural adviser to the Ministry of Agriculture was a certain Sir William Gavin, who also happened to work for the chemical company ICI. Many years later I talked privately to one of the country's most senior politicians, who had himself spent a couple of years at the Ministry, and he told me how disgraceful he thought it was that such a 'cosy relationship' had existed for so long between the agrochemical industry and the ministry that was supposed to exist to protect the nation's food supply. 'So why didn't you blow the whistle?' I asked him. He didn't bother to answer. He just gave me one of those looks that says: 'Don't be so bloody naïve!' I took the point. It had been made to me many times before (always in private, of course) by politicians who had served in what is now known as MAFF – the Ministry of Agriculture, Fisheries and Food. Interestingly, the 'Food' bit was not added until 1955 but there has never been much doubt about whose interests the Ministry was meant to serve: the farmers. They knew it, the politicians knew it and the civil servants knew it. Virtually every decision taken at the Ministry over the years, at least until the great food scares of the nineties, proved it. And virtually every politician I have talked to who has served for any time in the Ministry concluded that the best thing to do with it was to close it down and let other ministries take over its duties. The exceptions were those ministers who were themselves big landowners. They thought the Ministry did a wonderful job – as indeed it did. For them.

At the end of the war MCPA went on the market as Methoxone. It was a great hit with the farmers who knew it as 'cornland cleaner'. When it was dusted onto certain kinds of weeds that plagued cereal growers it made them twist, contort and die; it was far more effective than anything they had used before. No wonder. This was a powerful chemical, a hormone weedkiller. The pesticides used before the war were derived from naturally occurring mineral and plant products. These new poisons were as different from them as an Austin Seven is from a Formula One racing car. They were created in the laboratory, their molecules manipulated, one atom substituted for another. They were synthetic, relatively cheap to make and very effective. There was another reason for their popularity. From 1947, when Parliament passed the Agriculture Act, cereal farmers knew that it was almost impossible to lose money on the crops they grew. Subsidies had been introduced that virtually guaranteed them a profit, however much they spent on chemicals to kill their weeds or anything else.

Within a year of Methoxone hitting the market ICI was working on a new way of spreading the chemical. Until then it had been a pretty laborious process. The farm labourers had to walk up and down the rows dusting the plants by hand. How much more efficient to be able to do it mechanically, with a machine pulled by a tractor. That, of course, meant developing a liquid form of the spray and that duly happened. So, within a few years of the war ending, the more adventurous farmers were able to sit comfortably on their tractors and drive slowly up and down the field while a great boom behind them sprayed every single plant in the field. It was efficient and effective and it was not long before almost every cereal crop in the land was being sprayed with a hormone weedkiller.

As Graham Harvey wrote, this was not merely an updated version of what had happened before. This was qualitatively different. A new type of farming system had been born 'no longer based on an essentially biological process, tried and tested by a thousand years of crop rotation and sound husbandry, but rather a factory process that merely happened to use land'. In short, the age of pesticide farming had arrived.

By the end of the century there were three hundred different chemicals approved for use on crops in Britain. Others had been banned or suspended from use. Every year new compounds were being invented and sprayed on everything: the land itself, many of the animals, the harvested crops in the grain stores, in our parks, our homes and our gardens. Most were used to kill things: from large pests such as rats and mice down to the tiniest of plant-eating insects, aphids, mites, fungi, whether in the field or the barn or the containers carrying them to the supermarkets. Some were used to halt growth, perhaps to stop potatoes from sprouting. We were all being exposed to pesticides throughout our lives: through residues left behind in the food we ate and the water we drank and through sprays carried in the wind.

In those very early days no one seemed to worry about any possible side effects. The chemicals did the job: they killed insects and weeds and that was that. It was to be several years before serious questions began to be raised about the profoundly disturbing impact they might be having on other things. On the environment, the rivers and lakes and coastal waters into which, inevitably, they ran. On the soil, where chemical residues gradually built up. On the birds and mammals that fed on the poisoned insects. On the food we ate. On our bodies and our brains.

It was the American biologist Rachel Carson in her book *Silent Spring* who first alerted many people to the damage being caused by the unrestricted use of toxic chemicals. Other books on the same theme had been written, but *Silent Spring* touched a national and then an international nerve. For anyone who cares about the countryside and what has been happening to our food it is one of the most important books to have been written in the past century. 'For the first time in history,' she wrote, 'every human being is now subjected to contact with dangerous chemicals from the moment of conception until death.'

Silent Spring was published in 1962. It earned her the respect of millions and the enmity of the increasingly powerful agrochemical industry. At first they patronised her; she was, after all, just a woman – and a 'childless spinster' at that. Then they attacked. The industry spent more than a quarter of a million dollars – a fortune in those days – on a massive propaganda effort. They rubbished everything to do with Carson and her science. Pesticides were not only vital for the future of agriculture, they said; they were also harmless. Some journalists and scientists were persuaded. She was a crank and an hysterical woman, they wrote. But she stuck to her guns and when she appeared on national television to defend herself the American public saw someone very different: calm, persuasive and well informed, always prepared to defend her claims with evidence and yet more evidence.

One year after the book was published the Scientific Advisory Committee set up by President John F. Kennedy issued a report. *Silent Spring* was vindicated. The report concluded with the remarkable admission that, until they read the book, people had generally been unaware of the toxicity of pesticides. All these years later we may smile indulgently at such naïvety. How could they possibly have thought otherwise? The answer is perfectly simple and perfectly understandable: they believed what the industry had told them. And the industry had told them only about the benefits and not about the dangers.

Yet the risk from pesticides was not the central point in her book. Carson made it clear right from the beginning that she was not opposed to the use of chemicals in agriculture, even though they were poisonous. Her main point was that everything in nature is related to everything else. If you produce vast quantities of synthetic toxins, spread them across millions of acres of farmland, use them in homes and in gardens, it is inevitable that many unforeseen consequences will arise from that. That is such a blindingly obvious piece of common sense you might wonder that it needed saying. But not only did it need saying then; it needs repeating just as strongly today. We may be using different kinds of chemicals today but the basic questions remain:

- Has intensive farming with chemicals harmed the environment and are the chemicals we use today still doing so?
- Are they potentially dangerous to human health?
- To what extent is our food contaminated with the residues of harmful chemicals?
- Is everything being done that can be done to protect us?

To answer the first question you need only take a stroll on a sunny spring afternoon in any part of Britain where there is intensive arable farming, where the same crop is sown in the autumn, year in, year out. You will understand very quickly why Rachel Carson called her book *Silent Spring*. You will understand, too, the truth of the claim that everything in nature is related to everything else.

Most farmers want to grow more crops and in order to do that they try to kill the weeds that compete with their young plants and kill the insects that feed on them. Who can argue with the logic of that? Most of us don't like weeds, after all. Or do we? If you think of the wretched bindweed that tries to choke everything in sight in your garden or the even more damaging and indestructible Japanese knotweed that grows as tall as a man and kills everything beneath it then, no, we don't. But now think of the brilliant poppy, brightening a field of corn with its occasional flash of red or the less extravagant blue of the cornflower. Think of the corn buttercup or corn parsley or corn marigolds. Roll the names of some of the less familiar weeds around your tongue and imagine the centuries during which they have survived in the cornfields of England: the Lamb's succory and thorow-wax, the red hemp-nettle and shepherd's needle. Surely we can agree, if only out of sentiment, that a countryside in which they no longer exist is the poorer for it. Many of them are now extinct or are going that way – poisoned by herbicides or unable to survive in soil saturated with chemical fertilisers or simply unable to cope with modern farming routine.

For us it may be no more than a matter of mild, nostalgic regret that they are disappearing. To entire species of birds and small mammals it is a matter of survival. Vast numbers of the birds that once made that spring walk such a joy have simply disappeared as the weeds vanished. Some species – such as the tree sparrow – were once abundant and are now approaching extinction. In the years since the boom sprayers began their conquest of the land and farmers changed their age-old practices, there has been a catastrophic decline in the numbers of bullfinch and song thrush, skylarks and linnets. The reasons are simple: birds eat insects or seeds. If the insects are killed as efficiently as modern pesticides allow, then the birds cannot feed themselves and their young. They either die out or they go elsewhere. If there are no weeds to produce seeds for the seed eaters, it is the same sad story. Once again, the sprays and modern farming practices have done their deadly work all too efficiently.

And there are other victims still. If the slugs and beetles and butterflies are killed off there is no food for the small mammals that feed on them. The tiny dormouse has long since disappeared from our fields. Now its bigger cousin, the field mouse, and the shrew and the vole are under growing pressure. They cannot survive where there is nothing to sustain them. And the same goes for the birds of prey – the owls and the hawks – that swoop on the mammals. In nature, everything depends on everything else. There really is no mystery to it. Intensive farming with toxic chemicals is disastrous for wildlife.

Some will argue, sensibly enough, that nature is in a permanent state of change and we can never be sure why one species thrives at one stage in history and another struggles. The only sensible response is to compare what happens on land farmed intensively in lowland areas with what happens on land farmed without man-made pesticides and chemical fertilisers. It's like comparing a desert with a rainforest. In almost all of the studies carried out in recent years there were many more species on the land where crops were rotated and chemicals not routinely used. Some reported five times as many wild plants in arable fields; fifty-seven per cent more species; several rare and declining wild arable species found only on organic farms; twenty-five per cent more birds at the fields' edges; forty-four per cent more in the field itself in autumn and winter; more than twice as many breeding skylarks. And so the list goes on and on, whether we are talking about spiders or butterflies, ladybirds or wild flowers. Not even the most partisan supporter of conventional, chemical farming will argue that his method is as good for the bio-diversity of our countryside as organic, mixed farming.

But, you may ask, what has this to do with the main theme of this book, which is about the effect of modern food production methods on our health? Surely what matters is the food itself, not the birds and the insects and the wild flowers. True, but there are indirect benefits to be gained from an environment rich in bio-diversity. Doctors will tell you that it is better to go for a walk than to sit in a car and better to walk in the countryside than in a street filled with traffic fumes. They can prove it by measuring what goes into our lungs. I would go further. It is better to walk through countryside where you will see wild flowers dotting the hedgerows and skylarks chasing each other above the fields and hear the matchless music of a song thrush than through countryside where the hedges have disappeared, where endless, vast fields of corn stretch virtually unbroken to the horizon and where the only sound you will hear is the dull roar of a vast tractor pulling a boom sprayer. One of those walks gives us infinitely more pleasure than the other and our senses are stimulated in a quite different way. Quite simply, it makes us feel good. No, it's not possible to prove that it's better for us, but if a sense of wellbeing is not conducive to good health then I don't know what is. And if the scientists say that is sentimental, romantic bunkum . . .well, let them.

Are pesticides potentially dangerous to human health?

Again, a simple answer: yes.

There is a cocktail of more than five hundred different chemicals in our bodies. None of them was being made a century ago. They get there from the air we breathe, the water we drink and the food we eat. Some of them are relatively harmless; others are not. Among those that cause the greatest problems are the ones that accumulate in our body fat and cannot be broken down and flushed out of our system. Some of them are still used in pesticides manufactured overseas and find their way into the food we import. Modern pesticides used in this country do not come into that category. If we swallow them in our food we usually excrete them within three days. But that does not necessarily mean they are harmless. The danger is that if we swallow enough of them on a regular basis they will have a damaging effect on our body's metabolism. They affect our immune system, disrupt our hormones and damage our nervous system, heart and lungs. Fetuses and tiny babies are most at risk.

The companies who manufacture and sell the pesticides have consistently assured us over the years that there is nothing to worry about. So, by and large, have most civil servants – especially those in the Ministry of Agriculture, Fisheries and Food – and politicians, when they are in government. So, for that matter, have most scientists – though by no means all. Some are very worried indeed. And even those who are the least concerned admit that there are great gaps in the research that would be needed if we were ever to be entirely satisfied. As for medical doctors, all those I have spoken to in the course of researching this book – or, at least, all those who have given the subject any serious consideration – admit to niggling worries at the very least. Some are deeply disturbed.

In its *Guide to Pesticides, Chemicals and Health*, the British Medical Association says: 'It is almost impossible for any member of the population to avoid daily exposure to very low levels of several different pesticides in food and water. Consequently there is concern about possible adverse effects on human health arising from continual long-term low-level exposure, that is the potential for chronic toxicity.' Many doctors will tell you that there is no such thing as a 'safe' level of pesticide since some of the most dangerous toxins may be stored in our bodies for years.

And there is another problem. As the BMA puts it, chronic effects are likely to become apparent only after prolonged exposure to a chemical. In the case of cancer this may be a period of several decades, which leads to difficulties in identifying the causative agent. The slower the onset of the effect, the more difficult it is to prove the particular cause.

There is another complication too. Take the tragic case of Thalidomide. The deformities the drug caused in babies after it was taken by pregnant women were so appalling that the link was made relatively quickly – with the help of some brilliant investigative journalism by the *Sunday Times*. But if the drug had caused a more common malformation it could have been a very different story indeed. If, instead of babies being born without arms or legs, they had been born with, say, cleft palates would they have been spotted? Probably not. We might still not have connected the cause with the effect. So if something is being introduced into the environment or into our diet that causes an increase in the incidence of a common condition, such as lung cancer or heart attacks, the chances of associating it with those diseases are virtually nil. It is only when the condition is rare that we can be sure of picking it up.

We are, inevitably, exposed to any number of causative factors in our daily lives. We are not only eating food that may or may not be polluted with small residues of dangerous pesticides; we may also be cooking it in a way that adds to the risks: charring meat over burning charcoal, perhaps. We are also breathing in air that is very likely to be polluted, perhaps from the stuff that pours out of the back end of a car or a smoky old diesel lorry or the chimney of a factory. We may smoke cigarettes or breathe in others' smoke. We may be exposed to mild radiation in any one of a number of ways: radon emitted in houses sitting on granite, for instance.

We will certainly have accumulated in our bodies varying amounts of PCBs and dioxins. Some of us have more than others but no one is entirely free of them. Dr Vyvyan Howard of Liverpool University, one of the country's leading experts on the effects of chemical toxins on humans, has edited a book on the health effects of waste disposal policies, which contains some alarming results. Dr Howard is particularly concerned with the effect on babies. The degree to which they are affected depends on how much they absorb, both through the placenta and when they are breastfeeding. Recent studies from Holland have shown that the most highly exposed babies have a four point loss on the IQ scale compared with the least exposed.

Dr Howard reports the work of Dr Gunilla Lindstrom, a Swedish chemist who measured dioxin levels in her own breastmilk as she suckled her first child. The levels of dioxin in her body fell by fifteen per cent per month over the first six months. The poisons had been literally sucked out of her by her baby. It is estimated that a baby that is breastfed for six months will receive up to sixteen per cent of its total lifetime exposure to this class of environmental pollutant. Most damage is done by what passes to the foetus in the womb. Nobody suggests women should stop breastfeeding – there are too many other benefits to the baby – but there is obviously an urgent need to reduce our exposure to these persistent poisons.

Above all, our genes may have dealt some of us a hand of cards at conception that hugely increases the odds of getting cancer. A woman whose mother and grandmother both had breast cancer is at far greater risk than one whose family history is free of the disease. This 'pharmacogenetic' effect is well recognised in the testing of medicines. There is hardly a medicine that has been made to which a few genetically predisposed people will not be exquisitely sensitive. The same goes for environmental pollutants. But when pesticides are being tested, this effect is almost totally ignored.

Another difficulty with finding the cause of many diseases is the impossibility of establishing with any accuracy the extent to which we may have been exposed to any particular suspect chemical many years ago. It's hard enough to remember what we ate last week, let alone ten or twenty years ago.

And if in the end we do get cancer, it is fiendishly difficult to prove that it has happened because of pesticide residues we may have consumed for twenty years as opposed to the 'natural' process of ageing. Common sense tells us that we must die of something. We can reasonably suspect a connection if a man who has smoked sixty cigarettes a day since he was a boy develops lung cancer, or if a twenty-stone man who sits on his backside all day has a heart attack, but few cases are as clear-cut as that. The best we can do in the case of exposure to pesticides is to look at what they are capable of, to consider the evidence that has been amassed over the years, and to ask whether the concerns being expressed by so many people are being taken seriously by the political, industrial and scientific establishments.

When the age of farming with pesticides began in the middle of the twentieth century the chemicals in greatest use belonged to a group called organochlorines. They included such pesticides as DDT, dieldrin and aldrin. They killed insects well enough, but they also killed the birds that ate the seeds treated with them. The result was 'Silent Spring'. The chemicals also created a phenomenon known as food chain poisoning, which happens when animals or birds higher up the food chain feed on lower animals. Research carried out in the waters off California, into which the chemicals had run, produced a frightening result. It showed how the poison became ever more concentrated the higher up the food chain it travelled.

In the oceans, plankton are at the bottom of the food chain. By the time they had absorbed the poison from the water its potency had multiplied 265 times. When the plankton were eaten by small fish the multiplying factor was 500. When the tiddlers were eaten by large fish the concentration in the fatty tissue of the larger fish shot up to an astonishing 75,000 times. And when a bird came along and snapped up the fish, the bio-concentration was multiplied even more. Small wonder, then, that the birds who ate the most fish – the western grebe – were almost eliminated. Something similar happened with eels in Britain and their predators, herons. We talk about throwing our rubbish away. In truth, there is nowhere to throw it.

On land it was the same. One poisoned mouse might not be enough to kill a large bird of prey, but the toxin from the mouse would accumulate in the bird's own body fat or in its tissues and organs. And there it would stay, building up as it ate more and more contaminated prey until eventually it, too, was killed. Even if the birds did not die, the shells of the eggs they laid might be so thin that they would break before they could be hatched. In the 1960s birds such as owls, sparrowhawks and falcons were almost wiped out in cereal-growing areas. The poisons had the same effect on large mammals such as foxes and badgers. Their numbers were drastically reduced.

At the top of the food chain is the human being. The effect on us of organochlorines is more difficult to measure, partly because scientists have never been absolutely sure why the chemicals work the way they do. A number of research projects have found – unsurprisingly – that it is children who are the most vulnerable. Dr Howard says: 'In the United States this vulnerability is recognised by the imposition of a 10-fold safety factor in setting standards where foetal and infant exposure is regarded as problematic. In Europe there is no such provision and this is to be deplored. Babies are not small adults; they have different vulnerabilities. During the process of forming the brain, immune system, reproductive and other systems, the levels of various hormones and other naturally occurring chemicals in the body are absolutely critical and can be operating in concentrations of low parts per billion. This is the sort of level at which some environmental pollutants are active in the body. While they may cause only minimal or reversible effects in adults, that cannot be assumed for infants. We need to develop a "child-centred" regulatory system for chemicals and there is little doubt that this will need to include the phasing out of persistent bioaccumulative organic chemicals.'

Organochlorines were good at what they were meant to do, but their side effects were devastating. And, worst of all, they were almost indestructible. The toxins could stay in the soil for many, many years. Fields have been found to be contaminated with as much as thirty-nine per cent of the toxins that had been sprayed on them seventeen years earlier. With every year that passed it became increasingly clear that organochlorines were too dangerous to be released with such recklessness into the environment. The United States banned their use and eventually, many years later, Britain did the same. Not that the manufacturers stopped making them. They continued to do so and some of them are being used in agriculture in countries that supply us with food to this day. Some of the food we eat still contains traces of DDT for instance. Nowadays it is mostly used for controlling malaria but still it finds its way into the food chain.

As the dangers of organochlorines were finally being recognised and they were being phased out in this country, another group of chemicals was taking their place: organophospates, generally known as OPs. One group of toxic chemicals was being replaced with another group that is even more toxic. Organophosphates do their deadly work by attacking the brain. They make it impossible for some synapses – the nerve endings – to pass signals from one nerve cell to another. The effect is that the billions of messages sent from the brain cannot jump across the junctions. If you are an insect – or if you are a human and are exposed to enough of the stuff – you die. Smaller doses might mean, for instance, being unable to concentrate or suffering paralysis or diseases of the heart and respiratory system.

Some doctors and scientists I have spoken to are concerned about the increase in the past thirty years or so in the number of children who are being treated for attention deficit disorder. That happens to be the period during which OPs came into great use. Since the scientists cannot prove any connection their work has never been published and, of course, it may simply be that they are wrong and that children have simply been getting naughtier. It is more difficult to explain away the increase in the number of children with asthma during that same period.

There has also been an extraordinary increase in the number of young people stricken with the crippling – and ultimately fatal – brain disease, Alzheimer's. At one time it was believed to affect only the elderly, but that is no longer the case. Now it is found in people as young as thirty who are otherwise perfectly healthy. Dr Richard Harvey, the director of research at the Alzheimer's Society, said in September 2000: 'The number of young people diagnosed has increased dramatically. It has doubled, or more than that, in two years.'

One possible explanation is that there is now a greater awareness of the disease than there once was but Dr Graham Stokes, the director of mental health at BUPA, raised other questions. 'Could there be some environmental factor – some toxins?' he asked. Some scientists believe the answer to that is yes. Beta-amyloid is a prion protein – the same as those with which we have become familiar through BSE and CJD – that accumulates in the brain and causes the damage that we recognise as Alzheimer's. Research has shown that pesticides can increase the toxicity of prion protein and dramatically lower the age at which a disease such as Alzheimer's manifests itself. Other work has been carried out that shows a significant link between the use of pesticides and the early onset of Parkinson's Disease.

The great advantage of organophosphates, we are assured, is that they do not stay in the soil for years and years. When they are exposed to the sun and the rain, they break down and lose their deadly qualities. Nor do they bio-accumulate; they do not remain in the body and gradually build up, unlike their cousins, organochlorines. But it is not necessary for the pesticide to remain in the body for the damage to continue. If the effects of it were left behind then that would have the same consequences, although clearly it would not be detectable in the same way. In any case, some scientists and toxicologists believe it is impossible to be certain that OPs inevitably break down as reliably as we are assured and that they never accumulate in the body.

In order for pesticides of one group of chemicals to be used on the farm or in the home, they must be combined with other chemicals. That can change their character. Similarly, if we use one chemical on the roses in the garden and then another to do something else an hour later, there may be an interaction between the two in our bodies. One of them may simply add to the effect of the other. But what worries scientists most is that it may have what they call a synergistic effect. Instead of any damage being caused as a result of the interaction between the chemicals themselves, it is more likely to be because the initial disruption to chemical processes caused by one pesticide then produces an unexpected effect after a second exposure to a different product.

The evidence that OPs are harmful to humans has been gradually building over the years. There are long-term effects from both acute high doses and exposure to low doses over long periods. One of those effects is called neuropathy, disease of the nerves. It leads to muscles becoming numb and getting weaker. It usually starts at the ends of the nerves – in fingers and toes – and works upwards.

As the BMA concluded in its report, the effects of pesticides on the nervous system are difficult to detect but they have no doubt that they do exist and that they have caused considerable suffering to people who work with them and are exposed to them. Some of the most compelling evidence of the dangers from OPs has come from sheep farmers, who were required by law to dip their sheep in the chemicals twice a year.

Now we have moved into yet another generation of chemicals. They are called pyrethroids and, yes, the message from the industry is the same: these poisons are not only less toxic than those that have gone before, but they are safer in every respect.

Nature is perfectly capable of producing its own poisons. It does so all the time. There are thousands of them, most of which are designed to kill any insect foolish enough to try to make a meal of them. Many of them are lethal to humans. Many of the medicines we swallow today have been derived, one way or another, from toxins produced by plants or bacteria or fungi. Digitalis is one of them – an important heart drug that comes from the ubiquitous foxglove. The chrysanthemum produces its own poison. It is called pyrethrum and it is a very effective insecticide. If you're struggling to protect the young vegetables in your garden and would prefer not to spray them with any poison, you can do worse than plant a few chrysanthemums around the place. The trouble is that to produce all the pyrethroid poisons used in farms and in gardens from the flower itself would probably be impossible. It would certainly be very expensive. Thus, as so often before, the chemists have stepped in and designed their own synthetic version.

As with OPs, pyrethroids attack the nervous system. Instead of cutting off the messages that cross the nerve endings they make the brain repeat the messages endlessly. The effect is the same: death to the insect. At least, that's what was meant to happen. Some of the earlier pyrethroids were so ineffectual that when the bugs were sprayed they fell over but then got up again and carried on as before – a bit like a drunk who slumps onto the bar after several stiff drinks and then wakes up and orders another one. They might have been a bit wobbly, but they were still hungry. Something had to be done to increase the toxicity. The trouble is – and this is a familiar tale – the more lethal a chemical is to an insect, the greater the risk it may present to the environment and to people.

Again, as with OPs the reassurance offered with pyrethroids was that they break down easily and so cannot stay in the soil or store up in the human body. Synthetic pyrethroids are designed in the laboratory so that, even though they are more stable than natural pyrethroids, when they are exposed to sunlight or taken into our bodies they lose their effectiveness. If they did not, then they would present the same problems as organochlorines and live in the soil – or, perhaps, our bodies – year after year. So, from the perspective of the environment or people exposed to the stuff, that's good. But not if you want to kill bugs. By the time a synthetic pyrethroid has left the nozzle of the sprayer and hit its target it is no longer potent – hence all those insects in the early days with nothing much worse than a nasty hangover. From the perspective of the farmer, that's bad news. So a way has to be found to ensure that it retains its potency. This is where something called a potentiator comes in. As the word implies, it is a chemical that forms a synergy with the pyrethroid so that it remains potent for long enough to do the job required. And here we go again – back down the old road of messing about with toxins with results that cannot be foreseen. If a small amount of pyrethroid gets into our body there should, in theory, be no great problem. It should be broken down by our liver and expelled. But what if the potentiator stops the enzymes in our livers from doing their vital work? Well, then we have a potential problem.

It is interesting to note that a pattern has been established over the years. In the earliest days of pesticide use there was virtually no thought given to their safety. Most people were so excited by the potential benefits that they looked no further. Then, when questions were raised, the chemical companies rushed to assure us that we had no reason to worry: people like Rachel Carson were simply ill-informed hysterics. But when Carson and others were vindicated and more evidence built up that was difficult to refute, the companies adopted different tactics. They simply switched from one form of pesticide to another. So, first we had organochlorines being phased out in countries like the United States and Britain and then the same thing began to happen with OPs.

Now we have the pyrethroids and, as with their predecessors, we are told by the manufacturers and the regulators that there are no problems. Not that the old chemicals have disappeared altogether. Far from it. Some OPs are still in use in this country and they are used routinely abroad, almost always in poor countries where there are precious few protections for the farm workers who have to use the stuff or even fewer in the way of regulations to govern its use. Even organochlorines such as DDT are still used as pesticides in some countries. A cynic might conclude that instead of the manufacturers saying: 'My God! This is nasty stuff and we'd better stop making it,' they have said: 'Sure, this is nasty stuff. But if someone still wants to buy it . . . well, who are we to stop them?' As a result, not only are we using pesticides in this country that may have serious implications for our long-term health and the environment, but we are also buying food from countries that use chemicals which have been banned here for years.

That same cynic might ponder something else. Imagine what would happen if somebody who had become ill with brain problems or asthma or cancer were ever successfully to link their illness to a particular pesticide. It could open the floodgates to a torrent of legal actions for compensation.

Of all the pesticides in use today in Britain, the ones that worry many doctors the most are known as endocrine disrupters. The endocrine (or hormone) system is vital to how we function. It controls the way we grow and reproduce and the different ways in which our bodies are maintained in a healthy condition. It is a complex system of chemical messages that trigger our organs to produce a particular response. It is the endocrine gland that secretes the hormones that make the testosterone go roaring away.

Every parent of grown-up boys knows what happens when a teenager's hormones 'kick in' and he discovers sex. That nice little boy, who enjoyed nothing better than a sleepy cuddle before going to bed, becomes a monster. His voice drops an octave or two. He grows spots and suddenly decides he hates almost everything – especially his parents. Assuming he acknowledges their existence at all, he may devote a large part of his teenage years to making their lives hell. His mother might as well stick red hot needles in his eyes as kiss him in public; the reaction would be approximately the same. Then he grows up, becomes a civilised human being again, and realises that his parents aren't so bad after all. All of that is pretty much par for the course – extremely annoying, requiring parents to behave with superhuman restraint – but nothing to worry about. It's when something happens to interfere with the endocrine system that we should start worrying.

Hormones do a great deal more than control the delivery of sperm and eggs. They cause cells to multiply and to develop new functions. It needs only tiny concentrations of hormones to send those messages around our body through the blood supply. The endocrine system is never more vital than when we are developing in our mothers' wombs. If we are exposed to endocrine-disrupting substances at that stage the effects can be serious. The American scientist, Professor Frederick Vomsaal, has shown that rats become more aggressive than siblings born at the same time if they are exposed to marginally higher levels of testosterone in the womb.

It is an extremely complex business and scientists acknowledge that they have a long way to go before they understand completely how natural hormones are produced and how they work, let alone precisely how endocrine disruption occurs.

But there is a growing body of evidence that many different chemicals – some that occur naturally and many produced synthetically – are involved. A number of health problems have been recorded which may be linked to endocrine disrupters.

It is known that if a baby is born with no thyroid gland it can develop severe brain damage called 'cretinism' if it is left untreated. The question some researchers are now asking is what happens if thyroid hormone levels are even marginally decreased. This has already been measured in relation to the level of PCBs and dioxins in the body by Professor Janne Koppe and her team in Holland. It is possible that the lower IQ to which I referred earlier may in some way be related to this but at present we simply do not know. There may also be increased susceptibility to cancer and other diseases. The sex hormones oestrogen and testosterone are essential for the normal development of our reproductive organs, but an imbalance may seriously affect sexual behaviour, causing, for example, low sperm counts in men and abnormally early sexual development in girls. There is strong evidence that this is already happening. Equally, there is no doubt about increases in both breast cancer and testicular cancer.

It is almost impossible to prove many of the other possible effects. If a teacher says she has noticed over the past twenty or thirty years that more and more children have difficulty concentrating or are becoming more aggressive many scientists will dismiss her worries: it's purely anecdotal, they will say, or maybe there are social reasons. The problem is that there is no base line with which to compare, no section of the population that has not been exposed, no control group.

But many worrying effects have been shown in laboratory tests on animals and some of them have been shown when humans have been exposed to high doses of industrial chemicals. There is also a wealth of evidence from the largest laboratory in the world: our rivers and oceans. For more than half a century – ever since endocrine-disrupting chemicals came into use – studies have shown how fish have become feminised as the chemicals have leached into the water.

We know much less about the effect on humans of being exposed to low levels of endocrine disrupters over a period of many years. But there is enough concern amongst scientists for the Royal Society to have stated, in a report published in June 2000: 'Despite the uncertainty, it is prudent to minimise exposure of humans, especially pregnant women, to endocrine-disrupting chemicals.'

Let's take a handful of pesticides in current use, either in this country or in countries from whom we buy food.

Chlorpyrifos is an organophosphate insecticide still used in the home as well as on the farm. In June 2000 the American government announced that they had new evidence of its danger to children's health. The Environmental Protection Agency said the research shows the chemical can cause brain damage to rats. The Agency's administrator Carol Browner said: 'It is clear that the time has come to take action to protect our children from exposure to this chemical.' Its use is now under review, but it has not been banned in this country. It is known to be toxic to the nervous system. It has also been shown to affect the thyroid system in ewes. Links with cancer are suspected and more research is being carried out. It has also been linked by the German Federal Environment Agency to male and female deformities. It is sprayed onto apples grown here and abroad.

Carbendazim is a common fungicide in this country. It is used on everything from Brussels sprouts to blackcurrants, tomatoes, strawberries, apples and cereals. It disrupts the production of sperm and damages testicular development in adult rats, partly through disrupting the assembly of cells in tissues. It can damage the development of the fetus in the womb.

Experiments have shown that exposing rats in the womb to carbendazim leads to deformities such as lack of eyes and hydrocephalus or 'water on the brain'.

Vinclozolin is also a fungicide, used on oilseed rape, apples, peas and beans. It causes 'anti-maleness' effects. Studies show that when male rats are exposed to low doses in the womb and shortly after birth they undergo a series of changes to their sex organs, including reduced sperm numbers. Young male rats showed delayed puberty. Vinclozolin is on a draft 'priority' list of endocrine-disrupting chemicals produced by the European Commission in the summer of 2000. No action has been taken.

Aldicarb is used to kill insects and nematodes on crops. It is applied to the soil and is taken up by the plant roots and circulates around the whole plant. It was used widely in Britain until the early eighties and is still used extensively on potatoes, as well as sugar beet, carrots and parsnips. It is a carbamate pesticide which acts as a nerve poison by disrupting nerve impulses. The World Health Organisation classifies it as 'extremely hazardous'. Researchers found that some animals died within five minutes of breathing in aldicarb dust. Friends of the Earth put it at the top of the list of their 'filthy four' pesticides and want it banned. It has already been banned in some other countries. There have been many cases reported of food poisoning as a result of aldicarb residues.

Lindane is a hormone-disrupting chemical which has also been linked to cancer. The European Union has finally agreed that it should be banned but the ban will not come into effect until 2002. In the meantime it may still be used on food and fodder crops in this country.

The companies who make these pesticides and the regulatory authorities who control their use say the chemicals in use today are far less dangerous than many of those used in the past. In some ways that is true but it is not the same as saying they are safe. It merely proves how many more risks we took only a few years ago. We need much more research and information before we can be sure what the extent of the risks we are taking today is.

Can we be sure that our food is free from the residues of harmful chemicals?

On the contrary, what we can be sure of is that many of the foods we eat every day are routinely contaminated with residues of the chemicals with which they have been treated. The contamination occurs either when the food is growing or when it is in store or when it is in transit. The test here is not so much whether there is any residue, but how much there is. According to the companies who sell the pesticides and the government authorities who regulate their use, the presence of residues is not, in itself, a cause for concern. There are legal limits for pesticides known as maximum residue levels (MRLs).

The residue figures are published by the Pesticides Residues Committee. The committee is appointed by the government and consists of government scientists and representatives of farmers' and consumers' organisations and retailers. Until recently its reports were published once a year, based on 2,300 samples of food either produced in Britain or imported. The samples are taken from different types of shops and supermarkets around the country. Between thirty and fifty different types of food are tested each year including fruit, vegetables, cereal-based products, meat and fish. Baby foods are also monitored every year. The latest surveys show that the number of fruit and vegetables containing pesticide residues has increased considerably. Traces of chemicals were found in forty-three per cent of the fresh produce tested. That was an increase of ten per cent over the previous year. In some cases the amounts were above the MRL; in others the chemicals found had been banned in this country.

So there is no question about pesticide residues in our food. They are there. And that takes us to another of my key questions:

Is everything being done that can be done to protect us?

The answer to this one is, quite simply, no. If we accept – as we must – that pesticides are dangerous substances which have the potential to cause us great harm, then there are a number of serious concerns that must be allayed. They include:

- The way new pesticides are tested before they are given approval.
- The procedures for testing food for pesticide residues.
- The long-term effects of pesticide residues on our health.
- The 'cocktail' effect of toxic chemicals: what happens when one chemical is used in combination with another and how we are affected when we are exposed to a mixture of different chemicals.

The history of regulation in this country is not reassuring.

It was not until 1986 when the Food and Environment Protection Act came into force that the use of pesticides was controlled by statutory regulation. It took another four years before all the different regulations were in place. Until then there had been a voluntary system, the product of a cosy relationship between the big farmers, the agrochemical companies and the Ministry of Agriculture, Fisheries and Food. That might have remained the case to this day if we were not members of the European Union and bound by its regulations.

It is, when you think about it, quite extraordinary. Here we were, embarking on a revolution in agriculture the like of which had never been seen before, and the Ministry responsible for farming behaved with what seems to have been total disregard for anything other than maximum production. Farmers were not merely allowed but positively encouraged to use highly toxic chemicals without anyone having the first idea of what they might do to the environment, or to the men who were applying them, or to their livestock or to the people who ate their produce. But ignorance, in this case, is not really a defence. There might not have been much knowledge in this area, but there were many wise people who suspected right from the beginning that the use of such chemicals as organochlorines might have serious effects. Sadly, their voices went largely unheeded.

In 1950 a committee of eminent scientists came into being, under the chairmanship of the much respected Lord Zuckerman, with the task of looking into this new science of pesticide manufacture and what it might mean for us all. The committee did a good job. Zuckerman warned that there were real dangers in these powerful new chemicals and said the government of the day should bring in new legislation to control the use of pesticides.

So what happened?

Public records released thirty years later showed that the Zuckerman recommendations were either ignored or amended by civil servants. The Ministry of Agriculture thought there was no need for statutory controls. They would be an 'unwarranted interference with the freedom of commercial concerns'. Instead we had the Pesticides Safety Precautions Scheme which operated on a voluntary basis. It was no more nor less than a gentlemen's agreement. The interests of the consumer came a very poor second to those of the farmers and the chemical industry.

In 1987 the Parliamentary Select Committee on Agriculture was sufficiently concerned to set up its own hearings. Here's part of what the chairman's report had to say: 'We are not convinced that MAFF has responded seriously enough to the need for a greater effort to be made in the area of reviewing the toxicological data for older pesticides, approved under conditions more lax than today.'

Over the years, in the face of a growing body of evidence of the dangers of pesticides, the Ministry was adamant that farmers and the agrochemical companies should police their own territory. We can judge how effective that approach has been from the way in which some of the nastiest chemicals have been dealt with by ministry officials. Even when the risks were accepted in other countries, Britain was slow to act.

The use of DDT, one of the earliest organochlorines, as a pesticide in agriculture was ringing alarm bells back in the sixties throughout the world. The Americans decided to ban it in 1971. Officials in this country either failed to register the alarms or decided they knew better. So instead of a ban there was a gradual phasing out and a 'voluntary' ban from 1974. It took another ten years for a complete ban. In the case of two other powerful organochlorines – dieldrin and aldrin – it was to take another five years before they were completely banned in this country, in spite of all the evidence against them.

Many scientists around the world had already become convinced of their potential harm. It turned out that they were right to be worried. In 1976 blood samples were taken from nearly eight thousand women in Copenhagen. Over the next seventeen years 268 of those women developed invasive breast cancer. Each woman with breast cancer was matched with two other, healthy, women who acted as 'controls'. Analysis of blood samples showed that women with dieldrin concentrations above the average had a doubled risk of breast cancer. When the findings were reported in the *Lancet* the scientists who conducted the research wrote: 'Dieldrin was associated with a significantly increased dose-related risk of breast cancer.' To make certain that the women with cancer were not already diseased when the blood samples were taken, all those who were diagnosed with breast cancer within five years of the sampling were excluded. The result of that, said the researchers, was to strengthen the case against dieldrin.

Procedures have been tightened up since those days and since the mid-eighties there has been a recognition of the risks. But Britain still seems to take a relaxed attitude compared to many other developed countries. There has been enormous concern about the effects of the pesticide lindane for decades. Its use was severely restricted back in 1982 by Israel because of suspected links with breast cancer. Six years later it was banned in Sweden for that very reason and also because of the ways it persists in the environment. New Zealand banned it in 1990 because of its environmental impact. In Britain it was put 'under review' in 1992. And then . . . nothing until July 2000 when Britain voted, with other European countries, for a ban at European level.

In 1998 the Pesticides Safety Directorate began a review of two whole groups of pesticides, organophosphates and carbamates, to 're-assess the risks'. There are thirty-seven active substances in those two groups under review. If the Precautionary Principle were working in the way that its signatories intended it seems unlikely that they would still be in use – any of them – but they are. Chlorpyrifos, one of the most notorious organophosphates, is on the list. The Advisory Committee on Pesticides is aware of its dangers and has recommended that it should be banned from household products because of those dangers. It has also questioned the existing safety levels for chlorpyrifos levels in our food. But they're still there.

There is a list of chemicals as long as your arm which were once used with the full approval of the Ministry, only to be withdrawn or suspended when they were shown to be too risky. Who knows how much damage has been caused to the environment or the long-term health of the nation as a result of that? How different might things have been had it not been for the cosy relationship between the Ministry and the farming industry?

By the end of the twentieth century the system had changed. Today all pesticides must be approved by at least three ministers, so to that extent there is some political accountability. They have to rely on advice from the Pesticides Safety Directorate at the Ministry. It is responsible for new active ingredients, new products and new uses of existing products on the farm. The Directorate evaluates and assesses all the information from pesticide companies and provides summaries of the information to the government's Advisory Committee on Pesticides. The committee then makes recommendations to ministers about whether or not a new active ingredient or product should be given approval for use. But first, of course, the new chemicals must be tested in the laboratory to see what effect they might have once they are unleashed on an unsuspecting public. That testing procedure is at the heart of it all. If the people who carry out the tests are not getting it right, for whatever reason, then there are great risks for all of us.

Try to imagine this scenario: a pharmaceutical company has developed a new drug. It has been shown to be brilliantly effective in its field. The drug developers subject it to years of toxicity testing, according to recognised protocols. Having achieved a clean bill of health at enormous expense, the company organises clinical trials. These too are without evidence of ill effects. The 'wonderdrug' is launched to great acclaim. Suppose that somebody who is treated with the drug then dies. And then somebody else. And then another. Then hundreds. All those who die are Japanese. Under increasing pressure the wonderdrug is recalled and the data reassessed. It is realised that not a single Japanese native was used in clinical trials.

You would conclude that that is a preposterous scenario, would you not? Clearly the methods used for the testing were flawed, you would think. They took no account of natural variation in populations and as such were incapable of picking up all undesirable consequences. Something that is declared totally safe to Caucasians may be toxic to some Japanese. The study design failed to take that into account.

But then you look more closely at the way the study was conducted and discover even greater flaws. Not only were no Japanese people used; the entire test population was albino. And they were all a similar age: between sixteen and eighteen. What's more, they were all of normal bodyweight and this was maintained by feeding a perfectly balanced diet. None of them was allowed to drink anything but water, and they were all kept in a climatically controlled hotel for the duration of the trial. Inspection of their ancestors revealed that they were related to each other.

It is now beginning to seem even more preposterous. Yet it is on precisely such population samples that the safety of novel compounds is assessed, at least with regard to animal toxicity testing. And you do not have to take my word for it. That scenario was created not by me, but by a woman who worked for years testing new compounds for chemical and pharmaceutical companies. It is based entirely on her own experiences as a toxicologist. Her name is Janie Axelrad and she became so concerned about the methods of research in common use among toxicologists in the industry that she left it to carry out research of her own into the effects on humans of pesticide use.

She was appalled to find that the protocols for toxicity testing have been relatively unchanged since testing became mandatory back in the late eighties. It is true that there has been an increase in the number of species of animal used. For some groups of chemical it is now necessary to use at least one non-rodent species. Reproduction studies have also become more routinely used. But, as Axelrad says, all the studies rely on data from standardised groups of animals and as such fail to address the problem of individuality.

'Animal toxicity tests,' she says, 'have evolved to produce results that are easy for scientists to interpret and reproduce. The same few species of animals are invariably used for every test, because scientists have huge banks of background data for them. Many of the animal species used are albinos, since pigment makes some parameters difficult to assess.

'In the majority of studies, animals are all housed together, with control of temperature, humidity and air changes. They are all fed a perfectly designed feed, given only water to drink and originate from the same breeders. They are all brought in at the same age range, same weight range, and are kept free from diseases. Nothing can affect the animals except the product being investigated.

'Unfortunately this totally unrealistic situation ensures that extrapolation to the human population is fundamentally impossible. Humans are all different. We vary in race, colour and age. Some of us are overweight. Others anorexic. Our age range can be more than a hundred years, and each of us is different as shown by our fingerprints. Many people worldwide are vegetarian, and others live on a diet of hamburgers. There are even some people who enjoy a glass of grapefruit juice.'

In short, the animal studies bear no relationship to the potential realities of the world and have been designed to produce nice curves on graphs, with nothing exceptional that needs to be explained. Toxicologists – says this woman who used to do the job – like an easy life, and homogenous populations provide this.

Many other toxicologists will say there are perfectly good reasons why the number of variables is kept to a minimum. If they were not, they say, they would not be able to see the wood for the trees. The number of different animals needed would be vast, the amount of extra work needed would be staggering and the costs would rocket. So, since the tests can't be carried out on humans, this is the next best thing. But when we are earnestly assured that pesticides have been tested and are 'safe' we should take it with a pinch of salt.

Axelrad described a conference on pesticide safety at which a leading expert on risk assessment showed a slide of himself standing beside the stack of paperwork generated in toxicity studies. The paperwork reached his chin, and the message was clear: because there was so much of it we could all sleep safely in our beds at night. But quantity does not equate to quality. The vast majority of pages in that pile contained the evidence – in painstaking detail – that everything in the toxicologists' power was done to ensure that all the animals were identical and subject to identical treatment. Take away all of that, and there would not have been a great deal left. In Axelrad's view, the very fact that all the animals are effectively identical is proof of weakness in the procedures rather than strength.

You don't have to be a scientist to understand that humans react differently to different substances. Sometimes the difference can be dramatic. Most of us can happily guzzle handfuls of salted peanuts without any side effects except having to drink more to quench our thirst. For some people a single peanut can be fatal. We also know that many drugs are more toxic to the very young or the very old. So why test them on animals of the same age and characteristics?

New medicines must, by law, be tested on humans in clinical trials before they are given a licence for general sale. That cannot, for obvious reasons, apply to pesticides. All the more reason, you might think, for as much variation as possible to be introduced to the testing procedures. For instance, the toxicity of the pesticide malathion is linked to the amount of protein in the diet: the lower the protein intake, the more toxic it is. Some vegetarians may have a lower level of protein. So will some children. The toxicologists have many good reasons for testing as they do, but they do not pretend the system is perfect. The alternatives would be staggeringly expensive and time-consuming. So there are real questions to be raised about the testing of new compounds and about the way those new chemicals are subsequently put into use.

The chemical that does the killing when it is sprayed is known as the 'active ingredient', but pesticides contain many other chemicals to make them more effective. They might be solvents or thickeners and bulking agents. They might be surfactants, which make the pesticides stick to plants more efficiently, or stabilising agents. They might be chemicals to increase the killing power of the active ingredient or they might be something as seemingly innocuous as colouring agents or something to make the poison taste bitter.

In the United States the Environmental Protection Agency has classified two hundred or so chemicals used as 'inert' ingredients as hazardous air and water pollutants. Twenty-one of them are either known to be carcinogenic or suspected of it and fourteen have been assessed as 'extremely hazardous'. Some are suspected of being hormone disrupters. For every active ingredient developed there are many different products sold to farmers, with different mixtures or 'formulations' of the chemicals in them. One powerful pyrethroid for killing insects is available in twenty-six different products and is sold by sixteen different marketing companies. Even if there were the will to police its use it would be next to impossible to keep track of it.

There has been remarkably little research conducted into the long-term effects of pesticide use on our health. One of the problems is the absence of reliable data because adequate records have not been kept. For instance, in some areas of the country there is a much higher incidence of one disease than another. In Lincolnshire rates of breast cancer are about forty per cent higher than the national average. It happens that they grow a lot of vegetables, especially potatoes, in Lincolnshire and have used a great deal of lindane over the years. A strong link has been made between lindane and breast cancer. But there is no way of proving any specific link in Lincolnshire because we do not know where and when the lindane was used. It would be a relatively simple matter to require users of powerful pesticides to keep records and report them to the Advisory Committee on Pesticides. Reporting is mandatory in some states of America. But it does not happen here. In its own report published in 2000 the Environment Agency was critical of the failings of monitoring pesticides.

Another problem is that judgements on the safety of pesticides are based on animal studies, which does not necessarily tell us all we need to know about any adverse effects on humans. Organophosphates can produce subtle changes in our brains and nervous systems which are difficult, if not impossible, to detect in animals. And, as we all know, every human is different. Some farmers exposed to OPs over a period of time were horribly affected; others showed no symptoms. It is, of course, possible to build various factors into testing procedures to allow for variations but, since we don't even know why some people are more susceptible to some chemicals than others, it can never be foolproof. It goes without saying that testing of this sort on humans can never be an option. Even on animals, there are some tests which may simply not be carried out at all. There is no legal requirement for tests on the nervous, immune or endocrine system. Organisations such as Friends of the Earth believe that is deplorable.

There is also concern over what is known as the 'cocktail effect'.

If we were foolish enough to swallow a dozen aspirin tablets all at once it would have one kind of effect on us. If we were to swallow a dozen different kinds of painkillers the effect might be quite different. You do not have to be a scientist to know that mixing two chemicals together can produce unexpected results. With so many different combinations there is a risk that any synergy between the different chemicals may change the potency and the behaviour of the pesticide, both within the compound itself and within our own bodies when the different chemicals are absorbed or eaten.

What if a child is brought up by parents who follow the advice of the Department of Health to all of us to eat at least five portions of fresh fruits or vegetables every day? Well, in that case it is reasonable to assume that the child may swallow a whole mixture of pesticide residues during the course of a week. Remember, there are residues in forty-three per cent of all fresh fruits and vegetables tested in 1999. Friends of the Earth has found that fifty-seven different pesticides have been detected in the fruits which make up a typical home-made fruit salad.

But even that does not tell the whole story of our exposure to chemicals. Every time the child walks down the street he will breathe in petrol fumes from cars or diesel fumes from buses and lorries. If he has the misfortune to live near a factory there will be fumes from the chimney, sometimes very nasty fumes indeed. Every day of our lives we come into contact with a vast range of chemicals in addition to pesticides. Nor do we escape the chemicals when we go home and close the door behind us.

Let us assume that this little boy has just reached the age where he can play on a computer. His parents buy him a smart new one. They may not be aware that there is a problem with new computers. As the VDU on a computer warms up it emits a chemical called triphenyl phosphate, which is used as a flame retardant. With a new computer the emission is much greater than with an older one. Swedish scientists have conducted experiments which show that more than half of the eighteen brands of computers they tested emitted significant amounts of the chemical, which happens to be one of the OP family. The emissions were much greater than ordinary background levels. Although they were far higher the first few times the VDU was switched on, the chemical did not disappear. After more than 180 hours of use – and that's a lot of time spent in front of a computer screen even for a small boy – chemical levels were still ten times above the background level. It is not only computer manufacturers who use these chemicals; they are routinely used in dozens of products at home and in the office. And there are many other chemicals routinely present in homes and offices that can have worrying effects. What a witch's brew modern plastics can be, and all of it untested with respect to human toxicology. There is growing concern among scientists who specialise in the quality of indoor air as to the effect on us. At the least they can give us headaches, blocked noses and skin allergies. Some researchers believe they can increase the risk of epilepsy. At the worst – and in conjunction with other chemicals – we do not know.

So here we have this poor little chap, sitting in front of his computer after a nice meal of pesticide residues, breathing in contaminated air containing yet more organophosphates to add to the cocktail of chemicals in his growing body. What effect is all this having on him? Once again, we do not know. And the reason we do not know is that the tests that might provide the answers have not been carried out. Research in this area suggests strongly that when pesticides and other chemicals are combined, their harmful effects can be greatly multiplied. Hence the phenomenon of synergy to which I referred earlier. And yet, when pesticides are approved for use, levels are calculated separately for each chemical. If more than one pesticide is present, it is assumed that the total effect will be the same as that of the individual pesticides added together. The reality is otherwise.

One of the most common pesticides is glyphosate. It has been in use for many years and in one of its older formulations a chemical called polyexythelene amine (POEA) was added to the mix. POEA is a powerful poison – more acutely toxic even than glyphosate. So you would expect that when you combine the two you would end up with something pretty potent. And so you do. But not only is the resulting mix more powerful than glyphosate, it is also more toxic than POEA.

The Royal Society is sufficiently worried about the way mixtures of compounds may affect our health that it has recommended this should become a priority area of research. The problem is, research costs money – a great deal of money. Most research in the field of agrochemicals is funded by the chemical companies themselves. They are more interested in discovering new compounds that will protect their share of the market and boost their share price than they are in research that will raise even more question marks about some of their products. The universities don't have the sort of money required to carry out public-interest research on the scale that's needed so most do what their industrial pay-masters are prepared to pay for.

Towards the end of 2000 the Food Standards Agency announced that it was setting up a working group to look into the cocktail effect of different chemicals in our food. I sat in the audience of a conference at which a senior official from the Agency proudly announced that the group will meet six times a year. A member of the audience pointed out that, since there are approximately four hundred different chemicals involved, the number of different 'cocktail' combinations was vast. How many did the official think the group would be able to get through at each of its meetings? The official did not know.

As the former chairman of the Pesticides Residue Committee, Ian Shaw, put it: 'Little is known about the toxicological interactions between pesticides and therefore we must turn our attention to foods more likely to contain multiple residues. There are areas where I would like to see an improvement.' Mr Shaw is not alone in that.

Chemical companies want to be loved. If people get really alarmed about something they put pressure on the politicians and demand that something must be done. Then the politicians face a dilemma. Ministers can always bring in new regulations and ban chemicals that have been producing big profits and generally make life difficult for the manufacturers. That sort of thing scores brownie points with the voters, but it upsets big business. Political parties need to keep big business sweet for all sorts of reasons – not least the generous donations that finance their election campaigns. But if they turn a deaf ear to the demands of the public and things begin to go wrong, they risk being thrown out of office. If the noise is loud enough and lasts long enough it is usually the public that wins. Public opinion, when it is truly aroused, can be unstoppable.

Big business has a range of weapons at its disposal to protect its interests. Multinational corporations usually employ a vast army of public relations and 'corporate affairs' people to persuade us that what they are doing is, unfailingly, in the interest of the human race as a whole. Sometimes it is, in which case their job is easy. Sometimes it is not, in which case the company may own up, withdraw the particular product from sale, and set about producing something else to sell. Or they may lie about it, as the big tobacco companies did for decades. Not only were cigarettes dangerous; many of the biggest bosses knew they were dangerous and they mounted an elaborate and unscrupulous campaign of lies and deception so that they could keep advertising and selling their poisonous wares. No matter that millions of people all over the world were suffering and dying from emphysema or cancer as a direct result of smoking; they kept the lie going. Smoking was cool, relaxing . . . and harmless. It took many years for the lies to be comprehensively exposed.

The tobacco barons at least had the defence that smoking was a matter of personal choice. No matter how misleading or downright untruthful was the advertising they pumped out the final decision was, ultimately, our own. We could protect our health against the evils of smoking by not doing it and refusing to share an office with someone who did. We have no choice when it comes to chemicals. They are in everything: the air we breathe, the water we drink and the food we eat.

Ultimately the argument is not about whether we need chemicals. We do. Modern life would be impossible without them. The argument is whether the industry has been regulated strictly enough in the past. It has not and there has been massive damage to the environment as a result. We are still paying the price for factories that have been allowed to pour too much filth into the air and water and incinerators that have added to the problem. As for agricultural chemicals, the manufacturers are able to point to the enormous increases in harvests that have been achieved over the past fifty years. It would not have been possible, they say, without their synthetic fertilisers and their synthetic pesticides, and besides, the harmful effects of agricultural chemicals have been greatly exaggerated . . .

Back in the early sixties we saw how Rachel Carson was savaged in the United States by the big multinational chemical corporations when she dared to blow the whistle on organochlorines such as DDT. They wanted to get on with their business of making and selling the stuff in spite of the dangers. Perhaps they genuinely believed that people like Carson had got it wrong and that she was, as they claimed, simply hysterical and ill informed. Or perhaps, like the tobacco companies, they really knew the risks all along and were simply lying. We shall probably never know. Either way, the public was worried and the politicians had to react. These are not always easy decisions for politicians to take. They have to tread a fine line between balancing risk and creating alarm.

Measuring risk is particularly difficult when it comes to food. If there is one chance in a million that we may get a seriously upset stomach by eating fresh fruit contaminated with pesticide residues it would clearly be crazy for a government to advise us to stop eating fresh fruit. The health benefits of eating lots of fresh fruit and vegetables are well known and outweigh such a small risk. So that's an easy one.

If, on the other hand, there is one chance in ten thousand that we would not only get an upset stomach but also run the long-term risk of something much more serious – heart disease or cancer, perhaps – then it is less easy. And the problem is that the calculations cannot be made with any sort of precision. As the British Medical Association has said, there are so many factors when it comes to long-term health that it is almost impossible to pin any one effect to any one cause and be certain you've got it right.

What the politicians can do is make sure that the risks

are as small as science and human nature can make them. Governments must put in place a regulatory system that reacts swiftly when serious doubts are raised. The system must be completely independent of all the vested interests. It must be entirely open to scrutiny. It must always put the public health interest before the commercial interest of the manufacturers. It must ensure that the information the public is given about the risks is presented clearly, comprehensively and without bias. It must be accountable. The system we have had in Britain for many years has fallen short in many of those respects. Take pesticide residues.

The thresholds for pesticide residues – the Maximum Residue Levels – are agreed internationally by the Codex Alimentarius Commission (CAC). The CAC is a joint body of the Food and Agriculture Organisation of the United Nations and the World Health Organisation. The MRLs are based on what is known as the acceptable daily intake, agreed by the WHO with what is meant to be a considerable margin of safety. Over the years the thresholds have been lowered as new evidence has come to light of harmful effects and many pesticides have been banned altogether by different countries.

In Britain the Working Party on Pesticide Residues (WPPR) was the government body responsible for residue testing. It was reconstituted in March 2000 as the Pesticides Residues Committee. When the Ministry of Agriculture announced the name change it proudly declared that the new-look committee would have an 'entirely independent membership'. You might wonder why it had taken so many years for that to happen. But there was to be no rush. Even after the announcement was made it was to be two years before 'independence'. Nothing happens in a hurry where this particular Ministry is concerned. For instance, the report on residues detected in 1999 was not made public by the Ministry until the end of September 2000 – a long time to put together a few figures, you might think.

It is not as if they had a vast number of food samples to deal with. In 1998 there were forty-eight tomato samples and 180 lettuce samples. The style of presentation is intriguing, too. The summary of the committee's report on its website is intended for public consumption. It is possible for anyone with more than a passing interest to obtain the full report by phoning a Ministry number given on the web. I tried. I was told that the report seemed to be out of print. This was one day after it had been published. I persisted and eventually I was sent a copy. It turned out to be a copy of the report for the previous year. Such is life at the Ministry.

If you were to rely on the summary you would have been greeted with the following introductory question: 'What were the main results in 1999?' And here is the first line of the answer: 'Pesticide residues were not detected in 71% of the food samples tested.' No doubt it was meant to reassure us, but that seems to me to miss the point.

If an airline produced a safety report with the headline 'Ninety-nine-point-nine per cent of our planes did not crash last year' I think I might be a little puzzled. I would point out to the airline that what I wanted to know above all else was not how many planes had arrived safely – I rather assumed they all had – but if any had crashed. Whether it is aircraft safety or food purity we are talking about, it seems to me that we are entitled to make the basic assumption that the planes will not crash and our food will be pure. Call me an old cynic if you will, but if a government body has the task of guarding our health by checking up on pesticides in food, the top line in its report should be how many samples were contaminated, not how many were found to be free of residues. It may seem a trivial point, but I make it because the whole tone of the document on the website was one of reassurance. In an area as controversial as this, there is an important distinction between straightforward information and potentially misleading reassurance.

The report goes on to say that the great majority of the residues detected 'were of no health concern'. That applies not only to those below the legal limits but to those above them as well. If that sounds extraordinary perhaps we should look for the explanation in the definition of an MRL. If it's not a safety limit, what is it? Here's what the Committee says: 'MRLs are the maximum amount of a pesticide which is expected in a food if a pesticide has been applied correctly to a crop.' This, surely, is straight out of *Alice in Wonderland*. Not only is there no safety limit, but it is expected that there will be residues of chemicals in our food even if the pesticide has been applied correctly. We should not, however, let any of this worry us. Or so we are told. Not even the residues that were higher than the legal limits.

The report tells us what would happen to a small child if he ate one of the pears contaminated with chlormequat and what would happen if someone ate one-third of a sweet red pepper contaminated with methamidophos. In each case the effect would be nothing more harmful than an upset stomach. But what if a small child ate a third of the pepper for lunch and then the contaminated pear immediately after lunch? Would his stomach be more than 'mildly' upset . . . twice as upset, perhaps? And if, instead of eating only one-third, he happened to like the way his mum cooked sweet peppers so much that he ate a whole one? How much pesticide residue does it take to turn a 'mild upset' into something more serious? The summary of the report on the internet does not tell us. It merely reassures us – on the basis of eating one third of a pepper – that there is nothing to worry about.

The other striking aspect, once again, is the assurance that there would be 'no other health effects' apart from that mild stomach upset in toddlers. As for all the other residues detected the reassurance was total: 'The great majority of the residues detected, including those above legal limits, were of no health concern.' It doesn't get much more sweeping than that. They may have been 'of no health concern' to the experts on this not yet fully independent committee. Clearly they are of concern to many other distinguished scientists, not least members of the Royal Society. They are concerned not only with the 'cocktail effect' which I looked at earlier, but also with the fact that there is no statutory screening for the possible effects of hormone-disrupting chemicals. They are also of great concern to organisations such as Friends of the Earth and the Soil Association. Certainly it is possible to dismiss pressure groups and environmentalists as mere propagandists with their own axes to grind, but it seems to me both patronising and misleading to state as a fact that chemical residues are of no health concern. The only thing we can be certain of is that we do not know enough to be certain.

Nor do we know enough about the effects of pesticide residues on small babies. Clearly they are more vulnerable than adults, and not only because their bodies are so small. In the first months of life a baby's brain is growing at an extraordinary speed – faster than it will ever grow again. Even small changes in its development can have an effect on a child for the rest of its life. The European Commission has recognised as much. In June 2000 a European Directive came into force which said that processed baby food should have no more than 0.01 mg of pesticide per kilogram of any particular food. The new regulation comes into effect in Britain in July 2002. We may therefore assume that anything above that figure is, in theory at any rate, deemed to be potentially harmful. But many baby foods contain three or four different kinds of vegetable or fruit, so the baby might get three or four times the maximum residue in one meal. And what about babies eating fresh fruit and vegetables as against processed food? They might end up getting well above the 0.01 level.

That worries people like Lizzie Vann, who set up the

Organix baby food company, the first to manufacture baby food purely from organically grown food. She told me: 'We simply don't know what to make of these new regulations.'

The crucial question is what effect those relatively high pesticide levels will have on the baby's development. The answer is that we do not know. Nor do we know about effects on the fetus. If a tiny baby is vulnerable, so is the fetus – possibly even more so. Pregnant women are advised strongly not to smoke and to drink little if any alcohol during those nine months. But even if they stick scrupulously to those guidelines there are other things over which they have little control. From the moment of conception babies are exposed to pesticides that can pass through the placenta. Many have been detected in newborn infants, including chlorpyrifos, the use of which is now 'under review'. It is one of those chemicals that attacks the nervous system through the brain. The mother's hormones may also be affected by the endocrine-disrupting chemicals and may, in turn, affect the unborn child. I say 'may' because, yet again, we do not know.

Because there are so many uncertainties and, over the years, so many worries you might think that the Ministry would bend over backwards to provide the public with as much information about how pesticides are approved as possible. Not so. Approval is given by the Pesticides Safety Directorate and this statement gives you some idea of their approach: 'As a general rule all information, correspondence and other documents concerning pesticide approval are treated as confidential and cannot be disclosed.' So much for the public's right to know.

That statement was made in 1997. Since then there have been some half-hearted attempts to show the Directorate in a slightly more forthcoming light, but it seems not to have amounted to much. In 2000 the Directorate announced plans to open up its procedures to greater public scrutiny. In future we would have access to information about meetings of the Advisory Committee. Since then the agendas have been made available and so have summaries of the meetings but not, at the time of writing, detailed records, even though they were promised.

Yes, it is possible for the public to inspect the information that has been supplied to the Directorate by the pesticide manufacturers when they seek approval for a new product. But there is a catch the size of a cornfield. You may do so only after the pesticide concerned has been given full approval. If an organisation such as Friends of the Earth tried, for instance, to stop approval being granted on the basis of the data submitted they would fail. Once full approval has been given it is possible to inspect the data – or, rather, some of it. But the authorities don't exactly make it easy.

The first thing you have to do is buy a copy of the evaluation from the Directorate. You must then – and I promise you I am not making this up – sign a form confirming that you have actually read it. Quite why anyone would buy it without wanting to read it is beyond me, but one has to hope they have their reasons.

Having read and signed you must then travel to York, to the offices of the Directorate. It seems they have yet to catch up with e-mail or even the penny post. Once you are physically present in the office the data will be produced – all except that which has been blacked out for reasons of 'commercial confidentiality'. Now, at last, you may read it. But you will not be left alone with this precious document. A member of the Directorate staff will be with you at all times. Then, suitably chaperoned, you will be able to make notes. You will, if you wish, be able to copy the wretched document out word for word in your best copper-plate. What you will not be able to do is photocopy it or scan it into a laptop. The goods news is that you are no longer forced to use a quill pen and bottle of Indian ink. It is, by any measure, farcical. Why the absurd restrictions? Once again, it seems, the answer is commercial confidentiality.

But think about that for a moment. If the Directorate is concerned to protect a large corporation from having its secrets stolen by another large corporation, how does any of this help? The juiciest material has been removed anyway. If a multinational company wants to copy every document in the Directorate it can well afford to pay someone to do it. The people who might have difficulty putting in that kind of effort are organisations that rely on charitable donations, such as Friends of the Earth. The Directorate seems more concerned with guarding commercial confidentiality than it is with ensuring the public's right to know. Freedom of information takes a back seat, just as it always has.

I began this chapter by writing about some of my own experiences as a farmer, which have helped inform some of my own views. I suggested that most farmers should not automatically be seen as the villains of the piece, even though enormous damage has been done by the way in which they have been allowed to use dangerous chemicals for two generations. They are less culpable than those who sold the chemicals on a dubious prospectus with false assurances. They are also less culpable than public officials who, over the years, have so disgracefully failed to regulate the use of agricultural chemicals to protect each and every one of us.

The damage to our health is difficult to quantify. We simply cannot say how many people would have been spared great suffering or even an early death from a wide variety of illnesses if the use of toxic chemicals had been more tightly regulated and vigorously policed. The only certainty is that we do not know enough. That was true half a century ago and remains true today.

The Environmental Protection Agency in the United States tried to establish the risk to humans of the 600 active ingredients used in the 45,000 products marketed in America. It has been able to prove that only thirty-seven of them can be reasonably described as harmless. In this country the British Medical Association says no chemical pesticides can ever be proved totally safe. In the sixteenth century the philosopher Paracelsus wrote: 'Everything is poison, nothing is poison; it is the dose which makes the poison.' The World Health Organisation takes the same view: 'The increasing use of pesticides has led to widespread concerns about their potential ill effects on human health. The situation is particularly worrying in view of the lack of reliable data on the long-term consequences of exposure to pesticides.'

At every turn in the road that has taken us to intensive farming we come up against that immovable block: we do not know enough. The research has not been done. The data has not been gathered and analysed to present a full and accurate picture. In the mid-eighties a House of Commons select committee of MPs expressed the concern that 'none of the government agencies involved with pesticides seems to have made any serious attempt to gather data on the chronic effects of pesticides on human health'. The Health and Safety Executive had told the committee that it collected data only on acute cases. It said: 'The known statistics on poisoning, ill health and disease in agriculture do not allow us to form any judgement on illness resulting from chronic exposure.'

The committee concluded: 'We find this lack of epidemiological research quite unsatisfactory and urge greater efforts to be made in this area by the responsible public authorities.'

If any of those responsible public bodies did, indeed, make greater efforts the results were not apparent to the British Medical Association almost a decade later. Its own report stated: 'While no causal link has been proven between pesticides and forms of cancer, nervous and allergic diseases and reproduction problems, there are serious doubts about the scientific validity of some of the studies that have been undertaken and there is no epidemiological evidence available for many pesticides.'

As if that were not damning enough, the report went on: 'In other words, we do not know whether many pesticides are harmful or not in day-to-day use. This lack of information clearly needs to be corrected [. . .] Given the uncertainty it is difficult to make categorical statements on the degree of risk posed by pesticides, particularly the risk presented to human health.'

And that remains the official view of the BMA to this day.

As for the view of successive governments, well, most politicians see it as their job to offer reassurance. Perhaps ministers genuinely believe there is nothing to worry about. Or perhaps – and this is my own experience – many of them nurture secret doubts but are afraid to express them lest they be accused of scaremongering and undermining the public's confidence in the food they eat. That there are some doubts is clear from official advice. The Chief Medical Officer, no less, tells us officially that peeling 'is always a sensible precaution when preparing fruit and vegetables for small children'. The Health Minister had no trouble accepting that advice, but that sits oddly with what another ministry – MAFF – has had to say about peeling vegetables and fruit. In a report which found there were residues of the pesticide aldicarb in potatoes that were above safety levels for small children, it said there was 'unlikely to be a significant loss of residues from main crop potatoes as a result of peeling'. That's perfectly true. Aldicarb is only one of many chemicals that are 'systemic'; the growing plant absorbs the chemical to protect it against pests. If there is a residue it cannot be peeled or washed away.

But odder still is why we should have even to think about protecting the health of ourselves and our children from such toxic chemicals in this way. There is an inherent contradiction. Either the food is safe and we need have no worry about it. Or it is not and we have to peel away the skin which, incidentally, often contains valuable fibre and vitamins. The Food Standards Agency says that advice on 'topping, tailing and peeling' carrots no longer applies, but we are recommended to wash fruit and vegetables before eating. They say that has nothing to do with pesticides, it's just to make sure they're clean.

It is not enough to declare that a new compound or a new process or a new procedure is 'safe' because it has been tested in the laboratory and no ill effects have been found. The history of food is littered with examples of things going horribly wrong that were not forecast by the scientists at the time they were approved. If there is one infallible law of science it is this: if something can go wrong, it will. Too often over the last half-century the entire population has been treated as guinea pigs, the whole countryside as one great laboratory.

The precautionary principle should be at the heart of every new development. The principle was approved at a conference in the United States in 1998. It says: 'When an activity raises threats of harm to human health or the environment, precautionary measures should be taken even if some cause and effect relationships are not fully established scientifically. Recognition of the precautionary principle includes taking action in the face of uncertainty; shifting burdens of proof to those who create risks; analysis of alternatives to potentially harmful activities; and participatory decision making.'

Politicians pay lip service to that wise principle, but if the prospect of enough profit comes in through the door, precaution often flies out of the window.

Science and technology can do wonderful things; they can

do terrible things too. Take the polio vaccine. When I was a child it was common to see youngsters limping around the playground, their legs in metal braces, facing a lifetime of never being able to run or jump or kick a ball. There were other polio victims who spent their young lives in what we called an iron lung because they could no longer breathe for themselves. Many died. The polio vaccine, a drop on a lump of sugar, put an end to that misery. But it is possible that western medical researchers working with an experimental oral polio vaccine in Africa in the fifties caused as great a tragedy as the successful vaccine prevented: Aids.

The theory is that Aids crossed from chimpanzees to humans when doctors used the animals' kidneys to prepare the experimental vaccine. Those who believe that explanation argue that the sites of most of the earliest known cases of Aids – in Congo, Rwanda and Burundi – coincided with the sites of mass vaccinations with the experimental vaccines. Many researchers dispute the theory, but the fact is we do not know and perhaps we never will know.

Sometimes those who are to blame for a particular tragedy can be clearly pinpointed, as was the case with Thalidomide. More often the truth eludes us – as with the origins of Aids – either because corporations with vast financial interests at stake do their damnedest to conceal it or because, by the time the horror has unfolded, the trail has gone cold.

All too often we do not discover the consequences of science or technology that has gone wrong until it is too late. So there is one question we must ask on top of all the obvious questions about safety and effectiveness. Every time a scientist emerges from the laboratory with a new creation and every time the boss of a big corporation tells us how it will change our lives when they put it on the market we must say: 'Fine, but do we need it?' In the case of great medical breakthroughs such as the development of antibiotics or life-saving vaccines the question answers itself. In the case of pesticides it does not. We should enforce that key requirement from the Precautionary Principle about 'shifting burdens of proof to those who create risk'.

The scientist must always research, always seek new compounds or new processes or better ways of doing things. His career and reputation may be on the line or – at the very least – his next research grant or even his next mortgage payment. The corporation may need what he produces; the first duty of a board of directors is to look after the interests of the shareholders and they do that by grabbing a larger slice of the market and keeping profits high. But do we need it? If there is even the slightest element of risk – and it is a very brave scientist who can declare any new compound or new procedure entirely safe under all circumstances – then we are entitled to say no. Unless, that is, they can demonstrate that society as a whole will reap great benefits. If more people had said no more often in the past fifty years the world would have been spared many of the ecological, environmental and human disasters from which we have suffered.

And if we say yes? Well, then we must keep a much more careful eye on what is going on and we must be prepared to call a halt if we spot any worrying signs. If we do not constantly monitor the effects of each and every new chemical or process we run grave risks. The history of the half-century since Rachel Carson first began researching *Silent Spring* has proved that beyond doubt – not just in her native America but in this country too.

Farming practices and food production techniques have changed more in those fifty years than they have at any time in the history of agriculture. We have reaped some dividends in increased harvests, but we have paid a price for them. It may be that we shall never know just how high a price. It remains to be seen whether we have learned the lessons of those years and what they tell us about where we should go from here.

The truth is, of course, that neither the politicians nor the officials who advise them can be entirely satisfied that there are no risks from the use of so many toxic chemicals in so many different ways on so many different foods. So they hedge their bets a little and, ultimately, hope for the best. If the long-term effects on our health prove damaging they cannot be held to account.

The damage done to the environment by industrial agriculture is plain for all to see. Intensive farming with chemicals, the manic desire to grow that extra ear of corn whatever the environmental cost, has destroyed many of our most precious habitats. The legacy we shall bequeath to our children has been sadly diminished. Not for them the pleasures of downland alive with wild flowers and insects, or the serenity of an ancient water meadow. Many will never hear the glorious sound of a song thrush in spring or see a skylark furiously protecting her territory from would-be invaders. Some of the damage can be repaired over time given the right policies. Some of the losses are permanent. Many of us shed tears of anger and frustration and know that, in most cases, it is too late.

Sooner or later, of course, we come to terms with the loss. We can, when all is said and done, learn to live without the beauty of the downland or an ancient copse or a water meadow. It is desperately sad and, I believe, a betrayal of future generations, although I suppose it is possible to argue that our children and their children will not mourn for something they have never known. But intensive agriculture threatens another aspect of our environment upon which the very survival of future generations depends. In the next chapter I want to look at what is happening in the soil beneath our feet.

THE WORLD BENEATH OUR FEET

Soil

Sir David Attenborough is one of the great television broadcasters of the age. He has the lot: a wonderful voice, a sense of fun, an encyclopedic knowledge of every subject he tackles, boundless energy and – above all – genuine, unfeigned enthusiasm. You can't help feeling that whatever he is doing at that precise moment is the most exciting thing he has ever done. If he is spying on yet another exotic bird or stalking yet another exotic beast you know he must have done or seen something like it a thousand times before, but for him it seems always to be the first time. And because he is fascinated, then we are too. He has been everywhere and seen everything. Correction. He has been almost everywhere. Not even the great Attenborough has managed to report directly from the least explored and least understood environment on the planet: the earth beneath our feet. This is the last frontier. Try to imagine the programme he could make if he could manage to shrink to one-billionth of his size.

There he is, his microscopic figure crouched behind what looks like a boulder but is actually a tiny speck of soil, spying on the most extraordinary colony of creatures that even he has ever witnessed. 'And here,' he would whisper, struggling to contain his excitement, 'in this fascinating subterranean universe, we can see the most tremendous battle for survival going on. Over there, grazing calmly on some appetising soil fungi, is one of the most remarkable creatures in the underground world. It is called a springtail and, by golly, it needs that powerful little tail. Because any minute now it's about to be attacked by some of its many predators – an army of mites. And here they come! But the springtail knows they're there . . . and he's off! What a leap! If he were the size of a human that jump would have taken him clear over the Empire State Building with no trouble at all.

'That strange little creature with his powerful tail is pretty important to the survival of all us human beings up there in the human world. By grazing on the fungi he's stimulating them to reproduce themselves and that's essential to protect the roots of plants against some pretty nasty root-eating nematodes. But the nematodes have their enemies, too. One of them makes a tiny lasso to trap its prey. There are scorpions here, too – but it's jolly difficult to spot one of the little blighters. They're called pseudoscorpions and they're even smaller than a springtail. Just as well, really, since they're savage little blighters.

'Yes, all these creatures and countless more inhabit an extraordinary kingdom with its own hierarchy, its own complex mix of species. For next week's programme I'm going to be travelling even further below the surface of the earth to shoot the first pictures of bacteria that manage to live at the most astonishing depth and can exist without oxygen. Wish me luck!'

We will, David, we will.

A fanciful description, certainly, of what a miniature Attenborough would make of the world beneath our soil but it would be impossible for him to exaggerate the complexity of the life down below and the importance of what it does for us. It has been calculated that in one teaspoon of healthy soil there can be a billion organisms of more than ten thousand different species. That is twice as many species as have been described since Aristotle began to classify biological organisms more than two thousand years ago. Even more extraordinary is the fact that we know almost nothing about them.

We have cracked the map of the human genome, sent men to the moon and even invented a gadget that can remove the commercials from television programmes before we watch them. Yet when it comes to the soil beneath our feet, the very stuff of life, we are breathtakingly ignorant. We know that life on this planet would cease to exist if those countless billions of microscopic creatures stopped performing their strange functions and yet we pay them little or no heed. That is profoundly short-sighted not only because of what we know about them, but also because of what we do not yet know.

The few scientists engaged in the study of soil reckon that they have identified only about five per cent of all the various forms of life beneath our feet. Contrast that state of ignorance with our knowledge of the rainforests and the massive world-wide campaign to try to protect them and exploit their botanical treasures. Every schoolchild knows that the forests perform some vital tasks. The leaves on their trees suck in some of the nastier things that pollute our atmosphere and breathe out life-giving oxygen. We make a great fuss when the loggers move in with their chainsaws and destroy in a minute what nature took a millennium to create. Environmentalists and pop stars protest and demonstrate and then the politicians climb on board the bandwagon of public protest. Quite right too. The forests are, after all, the lungs of the earth and we need them.

We need them for other reasons too. In the war against disease it is astonishing how often the most powerful cure has come from the most humble plant. The greatest drug of them all is now acknowledged to be the most familiar – aspirin. No longer is it seen as just another painkiller, useful after a night out on the tiles. In the past few years we have begun to discover what else it can do: fight heart disease and even, in a modified form, cancer. And this superdrug was originally found in willow bark and leaves around 400 BC.

So many of the medicines on which we depend are derived from the plants of the rainforests and scientists are endlessly searching for something new. Might this strange little leaf or the bark on this exotic tree or the root of this colourful plant yield a cure for some terrible disease that has plagued us for so long? It might have – if it had not become extinct. How disgraceful, we say, that we are destroying so many species, wiping from the face of the earth for ever one variety of plant after another, before we have even discovered what life-saving secrets they may hold. We all know what has been happening and we are all shocked. We send off a few pounds to Greenpeace and we sign the latest petition. When we buy our new kitchen table we might even check to see whether the timber comes from a 'sustainable' forestry development. All very commendable. We may be fighting a losing battle – the greed of the logging companies is a mighty powerful force – but at least we are making an effort. And while the forest is still there researchers are doing their damnedest to unlock the secrets of its myriad plants for future research. Yet what do we do about the equivalent of the rainforest that lives in the soil? Virtually nothing.

The reasons for this sad state of affairs are not hard to find. Rainforests are not only ecologically vital, they are also beautiful. Who can fail to be moved by the sight of a magnificent hardwood tree brought crashing to the ground? Which of us is not roused to anger by the wasteland of stumps and smoking piles of ash left behind when the loggers move on? The great poets have vied with each other to hymn the praises of trees. When Wordsworth writes of 'a brotherhood of venerable trees' we can identify instantly with his sentiments. But try doing the same with a clod of earth.

The problem with earth is that at first glance it's pretty boring stuff; it takes a powerful microscope to bring its smaller inhabitants to life. We talk of the 'dirt beneath our feet'. Even the genius of a Wordsworth would struggle to find romance in a spadeful of soil. And there is another problem. We take the soil for granted. It has always been there and we assume it always will be there. There is no equivalent of a multinational logging company ruthlessly harvesting its riches to sell at a profit. We assume it is indestructible. We are wrong. In truth, it is all too fragile. And it is anything but boring. I might have allowed my imagination to run away with me when I conjured up a micro-sized Attenborough, but the activity in a spoonful of healthy soil is every bit as fascinating as I described.

The story of all that activity – at least, as much of it as we can understand – can be told without too much technical jargon. Its job is to break down complex organic matter into very simple pieces that are available to be taken up by the growing plant. Soil scientists split the organisms involved into three main groups: the macrobiological, the mesobiological and the microbiological.

One form of life at the macro end of the scale is animals. We eat our food and the food is digested and excreted. So, think of what happens to a cowpat on a field and you get a rough idea of the start of the process. When the cow has moved on a whole army of creatures takes over: earthworms and mites and snails and small arthropods and tinier worms called nematodes. They make a second meal of what has already been eaten once and break it down into much tinier particles with relatively large surface areas. Then they mix the organic matter with minerals in the soil and with fungi and bacteria. Crucially, they also take it below the soil surface – and that's where it gets eaten for the third time. That's where the real action begins.

It is at the microbiological level that the majority of the

organisms live and interact. More than half of them will be different kinds of bacteria. It's a pity about bacteria; they tend to get a pretty rotten press most of the time. We usually associate them with upset stomachs and think of them as 'germs'. The truth is we wouldn't last five minutes without them; in the great scheme of things nothing is more important.

Bacteria are incredibly varied in their size and shape and in what they do. Some of them make it possible for the soil structure to exist. They release a sugary gum that binds them to soil particles and then the soil particles bind to each other. No bacteria means no clumps of earth.

They also play an important part in chemically breaking down molecules passed down through the soil from that cowpat so they can be taken up by the plant. Scientists who study these things wax lyrical about how the process is carried out by very specific enzymes secreted by the bacteria which will 'chop up' molecules at specific places.

The other really important group of micro-organisms is the fungi. They also secrete specific enzymes that chop up molecules. It might seem a bit wasteful having different organisms doing the same job, but it is not. It's not like a butcher chopping up a chunk of meat into random sizes and selling it to anyone who wants to buy. It's more like a team of butchers working together to provide a specific cut of meat for each guest at a dinner party. Each butcher knows precisely where to cut a carcass to produce the right sort of steak or chop; so do the fungi and the bacteria. The difference is one of scale.

If all the different fungi and bacteria are not present in the soil then the guest or, in this case, the plant goes hungry. It does not get a full and balanced diet of nutrients. Not that the organisms are doing it for the sake of the plant; they're doing it for themselves. They have to, for the sake of their own growth and proliferation. When the fungi and bacteria die and decay they release minerals locked up in their cells to the benefit of the plants that are rooted in the soil.

There is another group of minuscule organisms helping that process. They're called protozoa and they roam the soil feeding on bacteria and fungi and each other. When they digest their prey they, like us, release the excess nitrogen which the plant gratefully accepts and takes in through its roots. But a healthy plant needs more than the fuel of nitrogen to force its way through the soil and up to the sunshine and stay healthy and strong. So do we when, eventually, we find it on our dinner table. It needs a wide range of minerals that are constantly being released through the weathering of soil particles.

Nature did a pretty inadequate job of designing plants when it comes to the uptake of minerals; they're not very good at it. But, as ever, she made up for the original design faults with some clever innovation. Most plants rely on a particular kind of fungi, known as vesicular arbuscular mycorrhyzae (VAM), which penetrates the cells of the plant's roots and creates a living bridge between the root and the minerals in the soil. It also helps the plant take up water. So the plant grows stronger and is able to fight off attack from pests and disease. In a healthy soil VAM account for about forty per cent of the fungi.

There are other organisms in the soil that help the plant fight disease. Not only do they give the plant the vigour it needs to withstand attack, they also form a physical barrier around the roots, making it more difficult for less friendly bugs to penetrate. Still others act like a needle stuck in a child's arm to vaccinate him against disease. The vaccine triggers an immune response so that when the measles virus attacks it can be fought off. In a plant something similar happens. The organism comes into contact with the root and triggers the response. That response is then logged in the plant's memory for when the nasty bug attacks. Other organisms produce chemicals that attack the bugs directly or even have an antibiotic effect on them.

And so it goes on and on, endlessly perpetuating itself, an unceasing cycle of activity involving vast armies of microscopically small creatures at every stage in this extraordinary subterranean kingdom.

All this is no more than the barest outline of one tiny part of the activity in healthy soil. It is infinitely more complicated than I have described or, if truth be told, than I am capable of understanding. Indeed even the most knowledgeable scientists in this field recognise that they know only a fraction of what there is to learn. But the crucial element in it all is that each tiny organism has its own distinct set of steps in this most complex waltz. Remove one species of organism and the danger is that the orchestra will lose its conductor; the rhythm and the efficiency of the whole system will be threatened. Remove enough and it may cease to function. For soil to be truly healthy they all have to work together.

Take our athletic little springtail, for instance. If there is not enough organic matter in the soil, the fungi on which he grazes will disappear and then the springtail himself will die out. No headlines will be written about his disappearance. The Worldwide Fund for Nature will launch no appeal to 'Save the springtail' as they do for the hump-back whale or the giant panda. He is neither majestic nor beautiful nor cuddly and his picture will appear on no poster. But the service he offers humanity is considerable. He is an important supplier of organic nitrogen which feeds the plants which feed us. It is yet more proof of the oldest law of nature: each function depends on each other. When everything works in harmony the results are truly impressive.

For hundreds of millions of years these tiny organisms have been going about their business, feeding plants and reprocessing material for the next generation of plants. But for the past half century we have launched a terrible assault on them, drenching the plants and the soil in which they grow with toxic chemicals and abandoning the farming practices that enable their underground activity to thrive. At first glance it seems we have managed to get away with it. The last fifty years have seen the biggest harvests produced in Britain since the first oxen drew the first plough. Drive through East Anglia or any of the corn-growing counties of England in the summer months and you will see vast fields of corn ripening, abundant and healthy. In autumn the grain silos will be stuffed full and the farmers will be preparing the fields to repeat the process that has served them so well over the years. How foolish, then, even to suggest that we are causing so much damage to the soil that we might one day live to regret it. The truth is, some farmers are already regretting it.

It is a crisp day in late autumn and Gary Coomber is ploughing a fifteen-acre field on his farm in Kent. Ploughing is one of those things that is wonderful to watch when it's being done by a professional and hellish to do yourself. I always thought it would be easy until I tried ploughing my own fields. The furrows would invariably end up looking more like waves breaking on a beach than railway lines disappearing into the distance. The trick is to keep your eye on a fixed point at the far end of the field and not deviate from that, but I'm afraid I never got the hang of it. Mr Coomber, by contrast, is a professional. He has been working this same field for many years and his furrows are as straight as the shaft of an arrow. But some things have changed. The first – and most noticeable – is that there are no birds following the plough. In the old days, he says, a flock of seagulls would swoop and call endlessly behind him, feeding off the rich harvest of worms in every sod of earth broken by the blade.

Now there are none because the worms have more or less disappeared.

The other difference is the tractor. This is a powerful beast – 100 hp – compared with what he once operated to pull the plough. That was a modest 35 hp – a moped compared with a Harley Davidson – but it did the job perfectly well. No farmer wants to use a bigger tractor than necessary; it costs much more and uses more diesel. But Mr Coomber has no choice. The earth no longer yields to the plough the way it did when the field was first ploughed. Where once the ridges of ploughed soil 'fell down like ashes', as Mr Coomber puts it, it's now more like ploughing through a field of boulders. He blames himself.

Like almost every farmer in this part of the world Mr Coomber had single-mindedly pursued one objective: to force every last pound of corn from the soil. He had done it by spreading synthetic fertilisers on the field and chemical pesticides and fungicides on his crops. It had been successful. His harvests were good and, in the early years, his cheques were large, not just from the sale of his corn but from the subsidies too. Then two things began to happen. The price the corn fetched on the market began to fall and so did the size of his harvests. As always in farming, there were a number of reasons for the declining yields. Weather is always a big factor; when farmers complain about the weather it is usually with good reason. But the land became increasingly unyielding, too.

When Mr Coomber began to farm these fields the soil was rich in organic matter and minerals. Like all the other land in the area it had once been a mixed farm: a few sheep and a few cows with the arable crops always rotated from one field to another as the years went by so that the fields that grew wheat or barley always had time to recover and regain their fertility before being expected to deliver another crop of grain. Instead of being sown with cereals year after year the fields would be sown with clover and grass to provide fodder for the animals and nitrogen for the soil, or beans, which also left the soil richer than they found it.

But when the politicians came knocking on his farm door with their offer of stupendous subsidies for every ton of grain he grew he threw out the old system and went hell-for-leather to grow as much wheat as he possibly could. Years later, when a crop of linseed attracted even better subsidies, he switched to that. Like many other farmers, he disliked the idea of farming for subsidies but it was a simple enough calculation: either you did it – or you went out of business. Today he is a chastened man, paying a price for feeding the soil with nothing but chemicals and killing off the life within it by endless applications of poisons to kill every bug and weed in sight. Now, he told me, the soil is not much more than blotting paper and something that keeps the crops in place. He intends to do things differently in future, to diversify as much as possible and to farm organically. He will grow less wheat but he hopes to get a higher price for what he does grow, assuming that the premium for organic produce remains at a high level.

His decision to begin farming without chemicals was taken for hard-headed economic reasons and so, ultimately, was his decision to switch to another system. He was reaching the stage where the cost of his inputs – the fertilisers and pesticides – was exceeding the value of his output. Like every other farmer with the same problem, he hopes to turn that around and return to profit. But there is something else too. Occasionally he tries to imagine what his grandfather would have made of the way the land was farmed for so many years. He does not like the thought. He wants the soil to return to the way it once was. He believes it can happen eventually – but it will take time.

From the southern end of England, let us now head north – about as far north as you can go without crossing the Scottish border. This is not arable country. It could scarcely be more different from the rich, forgiving farmland of the south where long summers warm the deep soil. The land in this hilly country is thin and sparse and usually acid, though it grows enough grass for the sheep to graze on as they have been doing for countless generations. And here, on the Cheviot hills, you will find a strange sight.

The land has been divided up into plots, twenty metres by twenty metres. Some of them are regularly sprayed with toxic pesticides or with sewage sludge and others are left alone. Some are coated with lime and others with synthetic fertilisers. This is known, in the jargon of the biologist's trade, as 'insulting' the earth. Some plots – even more bizarrely – have fine meshes embedded in the soil with electricity coursing through the wires. This is of course an experiment – the only one of its kind in Britain and one of the few in the world – to find out a bit more about what is happening in the soil.

The experiment – the Soil Biodiversity Programme – is being funded by the government to the tune of £6 million and it is being run by Professor Phil Ineson, a biologist from the University of York who spends his life trying to uncover the secrets of the soil. Professor Ineson is no woolly idealist with a romantic attachment to an environmentalist cause. He is a hard-headed scientist who believes only what he can prove in his laboratories by experiment or by years of observation in the field. And he is a worried man. He knows – not from fanciful imaginings but from what he has seen through his microscope over the years – that there is as much bio-diversity under the soles of each of his boots when he walks these hills as there is in a million acres of tropical rainforest. He wants to find out what it all does and how we can benefit from it for generations to come. What worries him is the possibility that, for some of the most productive areas of Britain, he may be too late.

This is a long-term experiment and, like any reputable scientist, Ineson will not pre-judge the outcomes. But some things have been established over the years. He can prove in the laboratory, for instance, what Mr Coomber has proved in his fifteen-acre field: that the regular use of synthetic chemicals above a certain level will kill off a vast number of micro-organisms in the soil and deplete the organic matter. You might expect that to be the case with toxic chemicals – pesticides and herbicides – but it is also true of fertilisers such as synthetic nitrogen. A plot that has not been sprayed with chemicals might have ten different species of bacteria; another that has been sprayed will have one. He also knows that certain organisms are vital for the plant to grow and thrive – what he calls the 'keystone' organisms. The trouble is, he does not know which they are. He and his colleagues have names for about ninety per cent of the vast number of different species. But as for what they all do . . . well, that's a different matter altogether.

He does, though, have an explanation for why so many of the tiny creatures seem to perform similar functions. Imagine, he says, some extreme conditions affecting the soil – perhaps several years of severe drought that may well kill off two out of three species of a particular micro-organism. When conditions return to normal the surviving species may enable the ecosystem to recover and all will be well. So this apparent duplication is actually nature's way of taking out an insurance policy. But what happens if you kill all the species in a given area? In the worst possible scenario there would be no recycling of organic matter and plants would not be able to release their nutrients back into the soil. The cycle that has been repeating itself for millions of years would begin to break down. Ultimately life on earth depends on that cycle. Without it there can be no vegetable growth. Without vegetable growth there would be famine on an unimaginable scale. Nobody suggests – least of all scientists such as Professor Ineson – that such an apocalyptic nightmare is on the point of happening. It may never happen. There is a vast number of soil types and one is as different from another as a human's fingerprints. Some are much more resilient than others.

But there is already strong evidence that the fertility of the soil in farming areas of many countries, including Britain, is being seriously damaged. Does that really matter so much? Won't science come to the rescue? There are, after all, other ways of growing food than relying on the earth to do what it has done for so long. Everyone knows about hydroponics and how it is possible to grow plants in water simply by adding the right mix of chemical fertilisers. Surely we could simply keep adding chemicals to the soil? Perhaps. But we would reach a stage where the manufacture and use of those chemicals would be unsustainable. On a very rough scale it takes three or four units of energy to produce one unit of food. And spreading chemicals on a field is not as efficient as you might think because the growing plants take up only a small proportion of them. Far more is washed away into ditches or streams or rivers, or soaks down into the watercourse. It causes pollution and it is terribly wasteful. That may be acceptable to the farmer so long as it is cheap to buy in relation to the profit he makes selling his crops, but when the profits begin to fall the sums look different. They look even less attractive if you add in the cost of cleaning up the pollution. Pressure for some sort of environmental tax on pesticides has been growing for some years now. There are signs that farming with chemicals may be approaching the stage of diminishing returns.

Dick Thompson is a scientist at the Soil Survey and Land Research Centre at Cranfield University in Bedfordshire. For some time he and his colleagues have been witnessing and recording changes in agricultural soils that could affect their productivity. Data from the Soil Survey's National Soil Inventory suggest that there has been a widespread loss of organic carbon in agricultural soils. Concentrations in arable soils have declined even further from levels that were already on the low side. More unexpectedly, those under grassland have also dropped over the last fifteen years. Soil organic matter is chemically and physically complex and the exact relationship between total soil organic matter and soil organisms and soil structure is complex and incompletely understood. But there is little doubt at the Soil Survey that depleted soil organic carbon equates in general with weaker structure. That means the soil can become solid and lifeless after activities such as winter harvesting of root crops. In simple terms, they appear dead. Damaged soil structure spells poor growth in following crops and hits farmers were it hurts most – in the bank balance.

Soil structure is extremely difficult to measure and describe in scientific terms but, like good art, you know good soil when you see it. It is friable, rich in organic matter, and breaks up easily into natural well formed crumbs. Poorly structured soil on the other hand is like potter's clay or paste when wet depending on its texture and turns to rock-hard bricks on drying. Dick Thompson's comments echo the concerns expressed by other scientists such as Professor Ineson: if you take the organic and living components out of soil then everything comes to a halt, just like cutting off the power to a factory.

Conventional systems of soil cultivation – ploughing, harrowing and rolling to create a seed bed – are also causing increasing soil erosion. The more the land is ploughed and worked, the greater the risk that it will simply wash away with the rain. If there is no structure to the soil, nothing to hold it together, no new organic matter to replace the old, then the effect is all the more dramatic.

Try taking a train ride through arable England in January or February. This is when the autumn-sown cereal and oilseed crops have been in the ground for several months but are still little more than small clumps with plenty of bare soil between them. You will notice at the bottom of the tractor lines, even on slightly sloping fields, that there are deposited deltas of eroded soil. Autumn seed beds become sealed at the surface as raindrops destroy the surface aggregates and create an impermeable skin to the soil. Until recently it was assumed that soil erosion is only a real threat where the soil is light and sandy or silty. This is where rills and gullies can form during heavy rain. Since much of lowland Britain has heavier soils nobody worried too much. But a detailed survey by Dr Tim Harrod and Dr Andy Fraser, colleagues of Dick Thompson's at the Soil Survey and Land Research Centre, has revealed that much greater amounts of fine soil are eroded from these heavier soils than from any other group of soils. That accounts for more than half the 2.2 million tons of soil Dr Harrod and Dr Fraser estimate are eroded each year in England and Wales.

It may not be very visible but it is immensely damaging. These fine sediments stay in suspension in land run-off and get into streams and rivers where they can destroy the spawning gravels of trout and salmon. The autumn floods of 2000 that caused so much heartache to so many people were muddy brown for a reason. They had fallen as rainfall on the fields of the upper catchments and gathered up topsoil as they found their way to the river network. So we are not only damaging the soil; much of it is literally disappearing before our very eyes.

Daniel Hillel, the Professor of Soil Physics at the University of Massachusetts and one of the world's leading authorities, wrote about the history of soil and civilisation in *Out of the Earth*. He described shocking examples of once-thriving regions of the world reduced to desolation by man's degradation of the soil. The southern part of Mesopotamia – the 'land between the rivers' – was once a great cradle of civilisation. In those ancients times there were 'fruitful fields and orchards, tended by enterprising irrigators whose very success inadvertently doomed their own land'. It is now part of Iraq and when Professor Hillel flew over it he saw 'wide stretches of barren, salt-encrusted terrain, criss-crossed with remnants of ancient irrigation canals'.

It has happened much closer to home over the millennia. A haunting example of soil abuse on a large scale, wrote the professor, is the Mediterranean region. It has borne the brunt of human activity more intensively and for a longer period than any other region on earth. For centuries rain-fed farming and grazing were practised on the hills of Israel, Lebanon, Greece, Cyprus, Crete, Italy, Sicily, Tunisia and eastern Spain. There was a mantle of fertile soil perhaps a metre deep. But no heed was paid to effective soil conservation. The land was denuded of its natural vegetative cover and the soil was raked off by the rains and swept down the valleys towards the sea. Today, where there was once good agricultural land, there are now rocky hills.

Professor Hillel speculates that the gradual loss of fertile soil was the reason why the Phoenicians, Greeks, Carthaginians and Romans were each, in turn, compelled to venture away from their own country and to establish far-flung colonies in pursuit of new productive land. 'The end came for each of these empires when it had become so dependent on faraway and unstable sources of supply that it could no longer maintain central control,' he writes.

It is easy to look at the disasters that befell ancient civilisations, smile indulgently and tell ourselves how much smarter we are nowadays. The truth is that with vastly greater knowledge we are making the same mistakes. The only real difference is that we are using the techniques of intensive agriculture to compress those mistakes into a much shorter time span. That doesn't sound very smart to me.

Nowhere is soil erosion more visible than in some of the loveliest areas of the land – the chalk downs of southern England.

When you fly low in a helicopter from east to west the appearance of the South Downs seems scarcely to have changed over the millennia. For mile after mile you follow an endless undulating sweep, the gentle contours broken by Devil's Dyke and Harrow Hill and Arundel Castle and a hundred small villages. To the south are the seaside resorts of Eastbourne and Brighton, Worthing and Littlehampton, and to the north the M4 carving its route from London to the Severn Bridge through the North Downs. Where they end you take a swoop south and Salisbury Plain stretches below you.

This is a land where men lived and worked so long ago that even the best informed archaeologists can only guess at many of their activities and marvel at their achievements. The burial mounds have left an indelible impression on the land, smoothed over with the earth of millennia but still visible from a thousand feet. The hill forts from the iron age still mark out their territory. If you come down to earth, cover your ears to blot out the roaring traffic from the A303 and gaze up at the towering monument of Stonehenge it is just possible, for a few romantic moments, to imagine that little has changed down the ages. Yet lower your eyes to the plain or lift them still higher to the skies and you see that everything has.

For centuries almost the only farmers on the downs and plains in this part of England were shepherds. They and their animals shared this land with a teeming population of butterflies and insects, birds and small mammals, wild flowers and herb-filled pastures. Now almost all of them have gone: killed off or scared off by the cereal farmers with their great boom sprayers and endless ploughing. In that respect the downs are no different from any other area where industrial agriculture has taken over and the precious ecosystem has been effectively destroyed.

Something else is disappearing too. This is not the deep, rich earth of East Anglia that can apparently take an indefinite amount of punishment from the incessant assault of the plough and the chemical sprayers. On most of the higher downs the mantle of soil is no more than a thin layer, fragile and delicate. It took an age for the soil to be laid down. It is vanishing before our eyes. Drive down the M4 towards South Wales and, as the downs slope to the road, you will see the white chalk breaking through where the soil has been washed away. The grass once bound it together but when the land is ploughed in the autumn and the bare soil exposed to the fierce winds and rain of winter it stands no chance.

I wonder how we shall explain to our children how and why we squandered such a precious heritage. Dick Thompson likens this process to the small change dropping from a rich man's pocket. But the difference is that a rich man keeps his money invested so that when he loses a little or spends some, there is always more cash coming in to restore his wealth. With the soil we are not investing; we are withdrawing. There is no guaranteed income and we are endlessly eating into the capital.

As Mr Thompson says, there tends to be an assumption in temperate climates such as ours that the soil will always be there and that it will always contain plenty of organic carbon. That familiarity and complacency has bred contempt. The lesson of history is that, even in temperate Europe, soil is all too vulnerable to foolish and greedy farming practices.

The extraordinary thing about the way we have treated the soil of this country since the middle of the last century is that so little attention and interest has been given to studying it. No public laboratory bears the word 'soil' in its name. No government agency is charged with its conservation or protection and it is only recently that the Environment Agency and the Department of the Environment have seen fit to allocate it as a specific policy responsibility, albeit on a minor level. As a nation we have spent millions on carefully cataloguing and protecting sites with distinctive wildlife, rocks or landforms – but not soils. Over the next decade, the UK will invest billions of pounds in the identification and clean-up of industrially contaminated sites that account, at best, for between one and two per cent of land. But what will we invest in the protection and sustainable management of soils beneath the remaining ninety-eight per cent?

The soil survey programmes of Great Britain once collected valuable information on the nature and quality of soils across our rich and varied landscapes. But all of them have been suspended for the last fifteen years. They were originally starved of funds by Mrs Thatcher and not one of them has been reactivated. In 1996 the Royal Commission on Environmental Pollution reported that there was serious cause for concern over what was happening to the soil. Its first recommendation was that there should be a national soil strategy. What did the politicians do? They made promises, as politicians always do. Then they did nothing.

In 1997 the new government promised that things would change. There would indeed be a national soil strategy and a draft paper would soon be published for consultation. When I began to write this book at the start of 2000 I asked the Ministry when I could expect to see a copy of the paper. I was told it would soon be on the minister's desk and would be published at Easter. In the summer I asked again. Still no paper. In the autumn I learned that the paper had indeed been delivered to the Minister but had ended up in the bin. Why?

It was deemed not to be good enough, not to make sensible and serious recommendations. It seems that this vital issue is considered to be of such low importance that the time, money and effort needed to produce a sensible strategy had simply not been made available. As things stand we spend tens of millions of pounds a year monitoring the quality of our air and our water but less than half a million on our soil.

If the last century has taught us anything about food production it is that we must, in future, always be prepared to think the unthinkable. It seemed unthinkable that the good roast beef of England could ever be seen as a health hazard, but it was. It seemed unthinkable that the rich earth of these fertile islands could ever be put at risk, but it has been. And what of the seas that wash our shores? When I was a child cod and chips was a cheap meal. Now cod has become almost a luxury food and soon it may be extinct. Unthinkable, but it has happened. We have the most extraordinary capacity for destroying what we value most.

FEAR OF FISH

Fish Farming

There are few spectacles in the natural world more awesome than a leaping salmon. Anyone who has seen it will tell you that it is not only a magnificent sight but intensely moving as well. The effort itself is heroic. The fish seems sometimes to balance on its tail, using every ounce of its strength to thrust itself up whatever obstacle lies in its path. Sometimes it will drive itself forward at astonishing speed and then, at the last moment, leap across the rocks lying in its path. Sometimes it will lie motionless, perhaps at the bottom of a waterfall, gaining strength for the task ahead. Then it will rise to a vertical position and thrash its way upwards as the torrent of water tries its damnedest to force it back down again. If it fails in its first attempt it will try again and again, its powerful tail thrashing from side to side, its whole body trembling and shuddering with the effort. Imagine a lone climber trying to scale a sheer rock face with only his feet to propel him upwards and an avalanche of snow trying to force him back down and you have some sense of the challenge.

You might think that such an achievement would be rewarded with some great prize. Instead, at the end of this extraordinary journey, lies the certainty of death. But not before the salmon completes the task for which the journey has been made. It is obeying some primitive instinct, perhaps a million years old, and returning to the water in which it was born so that it can give life to the next generation of salmon.

The salmon's whole life is a preparation for these few days of struggle. From the moment it is spawned the odds against it surviving to return to its ancestral home are many thousands to one. First of all its mother lays thousands of minuscule eggs in the gravel of a freshwater stream. Only a few of those eggs, fertilised by the milt of the male salmon, survive the first months of life. If the flow of fresh water does not deliver enough oxygen, if the temperature is not quite right, if the gravel shifts and they are exposed to one of a hundred hungry predators, they perish. Few make it to the next stage. Those that do survive develop into embryos, absorb the nutrients from the yolk of their eggs and eventually grow into tiny fry. Now the risks are even greater. They either learn the skill of dodging a hundred chasing mouths and hiding in dark corners or they die.

The few survivors spend the next couple of years growing into sizeable young salmon, called smolt. Now here is the truly extraordinary thing. Up to this point the salmon is a freshwater fish, but now, for reasons we cannot even begin to guess at, the salmon chooses to transform itself and begin the incredible journey which will take it many thousands of miles into the alien salt waters of the Atlantic. If you move any other freshwater fish into the sea it will die. As will an ocean fish transported to a river. But the salmon and its cousin, the sea trout, are like no other.

The transformation is not an easy one. First it has to reach the sea and that means leaving the relatively tranquil waters of its lake or loch and swimming many miles through a river, sometimes in full spate, sometimes when there is barely enough water in it for the salmon to swim. It will have to negotiate rapids and waterfalls and the risk of a violent death is considerable, smashed into rocks by the angry waters. And then, at the end of all that, there is the daunting prospect of adapting to its new saltwater home. Again, we cannot imagine the shock to the salmon's system. Those who have seen the smolts in their first hours and days in this foreign environment say that many appear stunned, floating on the surface like dead fish while they try to regain their strength, once again prey to a host of enemies. But if they survive that, then once again they are in their natural element, speeding through the waters with grace and strength, their mighty tails propelling them to new feeding grounds and, inevitably, new dangers.

Some will stay for only a year in the ice cold waters of the Atlantic before they begin their return journey. Others – again for reasons we cannot fully understand – will stay longer. By the time that primitive instinct drives them to find the mouth of their own river some are monsters of twenty pounds or more. Most are less than half that. But all strive to fight their way up the river to the water in which they were spawned or die in the attempt. Precisely how they find their way home is still a mystery, but it is believed that the salmon can detect water flow and temperature change and also, bizarrely, use their sense of smell. It is said that a salmon's sense of smell is one thousand times more acute than a dog's.

Once they are back in fresh water they never feed again. Every fibre of their being is now concentrated on finding the exact spot where their parents laid and fertilised their eggs so that they can do the same. When the female finds it she digs a hole in the gravel, lays her eggs and waits for a male to come along and fertilise them. Then she covers them over with a flick of her tail. What a task, and what a time to be starting that final journey. After swimming thousands of miles the salmon are exhausted, but they are driven by the oldest urge of all – for the survival of the species. It is hardly surprising, then, that the sight of a leaping salmon inspires such awe.

*

The last time I saw salmon leaping from the water I was moved not by admiration of the wonder of nature, but by anger at the greed and stupidity of man. The salmon were not battling against natural forces in their own element; they were in cages in a Scottish loch, protected against every species save the most predatory of all: humans. They were spawned and hatched in captivity, bred and fattened in captivity and will die in captivity. For the last weeks of their lives they will be starved of food, to purge them of the chemicals they have been fed. They will never know the freedom of a freshwater river or vast ocean, nor feast on the varied diet of wild creatures that give them such a distinct and subtle flavour.

These salmon were leaping, not for the eventual fulfilment of some primitive destiny, but because they were crammed, countless thousands of them, in their floating prisons, frequently tormented by sea lice and disease and kept alive with highly processed food and endless chemicals. This is intensive fish farming, a phenomenon of recent years which is not only destroying the ecology of some of the most precious waters on earth but is also threatening the future of our wild salmon. It is also, I believe, adding to our own health risks. Like so many other forays into new forms of food production, it seemed a good idea at the time.

A couple of centuries ago fresh, wild salmon was so abundant and so cheap that it was the staple food of the poor. Apprentices in London went on strike in the eighteenth century over their demand that they be fed salmon no more than five times a week. But over the years things changed. The great canneries went into operation. In Alaska millions of salmon, instead of being hauled from the water at the end of a rod or in an Eskimo's net, were trapped by commercial operators by the ton and ended up in tins on grocers' shelves. Those that survived were the treasured prey of amateur anglers or professional poachers. In Scotland generations of well-heeled gents were happy to spend small fortunes to stand all day in icy water flicking artificial flies into deep pools and waiting for the catch about which they would be able to boast for years to come. The salmon became the nation's most prized fish. I can remember my own father, who was born to a working-class family before the First World War, telling me that he had never once eaten fresh salmon. It was the ultimate luxury and not for the likes of him and his family. As for smoked salmon, that was the last word in delicacies.

So here was a food that had everything. It tasted wonderful, did you a power of good and had an exclusive cachet. It could only be a matter of time before someone, somewhere, found a way of exploiting it for profit. And so it was: it happened in the waters off the west coast of Scotland in the late sixties. Not that the Scots were the first to experiment with aquaculture. The Japanese had been involved in it for many hundreds of years. In medieval times the monks farmed carp. But twentieth-century, intensive aquaculture owed little to that.

The big argument in favour was that it would take the pressure off wild fish stocks. Many people, desperate to ensure the survival of wild Atlantic salmon, wanted badly to believe that. They were to be horribly disillusioned. The exact opposite was to happen. What they sought to preserve they ended up damaging – perhaps destroying.

By the middle of the seventies salmon farming looked like developing into a promising cottage industry, helping to create local jobs for the men who once made their living as crofters. By 1980 they were producing about 800 tons of fish. Twenty years later it was no longer a cottage industry but an international business. The stranglehold of the big corporations is now so tight that all but a handful of local fish farmers and crofters have been forced out of business or have sold out to a handful of multinationals who control three-quarters of all the farms.

By the end of the century the salmon cages were producing nearly 130,000 tons a year and the industry is still growing. As the farms have become bigger and more intensive, so the number of local people employed by them has fallen. What had once been an industry in which hundreds of individual fishermen and crofters and gamekeepers and poachers had made modest livings in one of the most beautiful and unpolluted environments on earth had now become big business. And no longer is the home of the wild salmon unpolluted.

If a chemical company builds a factory on the edge of a town the pollution may be there for all to see. The plant itself invariably dominates the landscape in all its ugly utility. Its by-products may come spewing out of the chimneys, sometimes coating nearby homes in dust, sometimes poisoning the earth for decades to come, sometimes even coating the lungs of children. Its owners may defend it on the basis that it provides work for local people and its products are essential in a modern society, but there is no denying its impact on the environment. With fish farms it is different. The pollution is less visible and more insidious – out of sight and, they hope, out of mind. But that depends on how determined you are to find out what is going on.

I first saw fish farms some years ago in the bays of the Isle of Skye, that magical island of savage mountains and gentle lochs that earned its place in history by offering a landing place to Bonnie Prince Charlie on his ill-fated mission to seize the crown. On a clear day, from the top of the mighty Cuillin range, the farms look like little more than empty crates bobbing on the surface of the waters – unsightly but relatively innocuous, like a spot on the face of an attractive young girl. You expect them to drift away with the tide and the current. But, unlike the blemish on a clear skin, they did not disappear with the passing of the years. And the closer you get the more brutal they become.

One warm summer evening I swam out to one with a friend who has a croft on Skye.

A pair of seals joined us for the company, though they disappeared when my friend tried to prove his theory that seals – like pilot whales – are supposed to be attracted by the human voice. He sang to them. All he proved is that seals are definitely not tone deaf. Apart from that it was a perfect evening. The water seemed, to my jaundiced London eyes, pretty clear. My friend, who has been swimming in those waters for nearly forty years, told me how they had changed: not just here on Skye but in lochs throughout the Highlands and Islands. Once the water truly was crystal clear; now it was tainted. Other nasty things were going on, too. Local newspapers had begun to report cases of swimmers being attacked by shoals of sea lice – always when a salmon farm happened to be in the vicinity.

As we swam, a gannet flew overhead and we watched her dive for her supper, dropping from the sky in a flash of streamlined feathers, scarcely breaking the water as she hunted her prey. I wondered what she made of the salmon in the cages.

A wild salmon is a beautiful creature. Its skin has a silvery sheen, its muscled flesh is firm and a delicately pale pink, its fins and tail designed for speed and agility. Its captive cousin is a sorry specimen by comparison. Compared with the king of fish, this is a lowly subject indeed in its squalid captivity. I don't know how many fish were in those cages; it was impossible to count. But the biggest and most intensive fish farms were then housing up to 40,000 in a cage. The most modern cages hold several times that number. In the worst cases it has been calculated that the fish have no more than the equivalent of half a bathtub each. Imagine two salmon living in a bath of water throughout their adult lives and you get some idea of their hellish existence.

The effect is entirely predictable. By their nature salmon

are aggressive fish. In the wild they must kill other fish and marine creatures to feed. Confined like this, they frequently attack each other, so their fins and tail are often damaged and their bodies scarred.

Close confinement also exposes the fish to attack from sea lice. Sea lice are as common on a salmon as fleas on a dog or cat. In its natural environment a salmon might pick up one or two lice in the course of its life but they will do no harm and eventually will disappear. When you pack fish together in cages you provide the perfect breeding ground for the lice. The water may contain millions of lice and the fish can become infested. It makes the life of the salmon hell.

For a farmed salmon to be profitable it must grow much more quickly than nature intended. All sorts of tricks are used. The hatchery tanks are covered over so that they are completely dark. Then bright lights are shone into them. The fish are fooled into thinking that they are passing rapidly through their natural seasons.

Salmon are carnivorous and they are at the top of their food chain. It's one thing to fatten up a steer to produce beef by giving it mountains of barley to eat. But salmon eat other fish. The comparison would have to be with farming lions or tigers – wild animals that hunt their prey – and we do not farm such animals. It is, in many ways, a biological nonsense. We catch the fish for them and then feed it to them.

And here's another problem: catching fish in, say, the South Atlantic and transporting them to Scotland is fiendishly expensive and, ultimately, not sustainable. So the producers are always looking for cheap food. The diet most farmed salmon get is abnormally high in oil. That gives them almost endless diarrhoea and softens their flesh. One result is the thick white lines in cheap salmon: it's fat.

Even without the diarrhoea there is an enormous amount of fish faeces suspended in their cages. That can lead to other infections, one of which attacks their gills and makes them produce an abnormal amount of mucus. Farmed salmon are also prone to infectious salmon anaemia (ISA), which is a sort of fish flu. In 1998 the disease swept through the salmon farms of Scotland. Four million fish were slaughtered and three-quarters of the salmon farming industry was 'quarantined'. The quarantine did not, as it was hoped, put an end to the disease.

There are many more illnesses: cardiomyopathy syndrome (CMS), otherwise known as deadly heart disease; infectious pancreatic necrosis (IPN). The list goes on. Intensive salmon farms are, by their very nature, unhealthy places for fish. They swim in their own faeces and use their home as an open sewer, so it's hardly surprising.

It is worrying that farmed fish suffer from so many diseases. It is even more worrying that those diseases spread to other species. One of the problems is that the cages do not make effective prisons. Hundreds of thousands of farmed salmon have escaped since the first fish farms were set up off the coast of Scotland. By the time you read these words the figure will almost certainly have passed the half million mark. Millions more have escaped off the coasts of Canada, Chile, Norway and the United States. An article in *Nature* magazine estimated that as much as forty per cent of Atlantic salmon caught by fishermen in areas of the North Atlantic Ocean are of farmed origin. Some rivers in Norway are now so infested with escaped fish that up to ninety per cent may be of farmed origin.

Good, you may say. The escaped salmon get a better life and will augment the stock of wild salmon. Sadly, it is precisely the opposite. These fish have not been bred to imitate their wild cousins. They are fatter, not fitter. If they breed with them – and they do – they cause immense harm to the gene pool. According to *Nature*, increasing evidence suggests that not only do farm escapees interbreed with Atlantic salmon they also may hybridise with, and alter the genetic make-up of, wild populations of brown trout. They also spread their lice to wild salmon. For years that obvious fact was denied by the industry and, indeed, by successive government ministers. One government scientist bravely stuck out his neck in the *New Scientist* in 1998 and said that the link was 'as plain as the nose on your face'. But he wrote the article anonymously. The fish farm lobby is a powerful one.

Over the years Scottish politicians have, with one or two exceptions, done everything they can to defend the fish farming industry. As a wild salmon responds to a juicy fly, so politicians respond to votes. Fish farms created jobs in an area where jobs were badly needed. Any politician seen to be hostile to the job creators might suffer at the hands of the voters. Result: the fish farms have been able to get away with murder. I have talked, in private, to a number of very senior Scottish politicians about the effect of the fish farms on the environment and on wild salmon stocks and, to a man, they have told me it is deeply worrying. So would they say that in public? They would not.

Perhaps there is a certain amount of embarrassment at the way in which the industry has been regulated. In its earliest days there were no controls at all. It was not until 1989 that the River Purification Boards were given any real powers to control pollution and the Scottish Environmental Protection Agency did not come into existence until seven years later. So, right from the beginning, the industry has had an easy ride. The Minister in the Scottish office who had responsibility for fish farming under the Conservative government took a new job after the government lost power. He became the chairman of the body representing the interests of the salmon farmers, the Scottish Salmon Growers' Association, since renamed Scottish Quality Salmon.

But in August of 1999 the Agency's area director, Professor David Mackay, went to Norway to present a paper at a conference on the environmental effects of the cages. Even allowing for the careful language of an academic expert at a conference of his peers it was devastating. If there was a watershed in policy on fish farming, then this was it: 'The case that damages to stocks of sea trout and wild salmon by sea lice associated with caged fish farming is very serious in certain circumstances has been made to the point that it should now be accepted as beyond reasonable doubt.'

The professor was saying no more than what a thousand anglers had been saying for years, but what mattered was that he had the authority of the Agency behind him. The minutes of the next SEPA board meeting stated that his paper had been 'a slight shift in policy' – another masterpiece of understatement. That phrase – 'accepted as beyond reasonable doubt' – was the killer. The stocks of wild salmon in the finest salmon rivers in Britain – some of the finest in the world – have been literally decimated. Every angler, every landowner, every local who has ever wielded a rod or cared for a stretch of water, tells the same story. I heard it over and over again when I went to the Highlands and Islands to research this book: the salmon have gone.

I stayed for a few days with a friend who lives in an ancient castle guarding the mouth of one of the loveliest of all the lochs. His family have owned land in the area for generations. This is poor farming land but it is rich in so much else. Much of the family's income over the years came from anglers from all over the world who had gone there to catch the famed Atlantic salmon. Visitors still come to walk the hills and climb the mountains and breathe the clean air, but they do not come to fish. It does not matter how skilled you may be with a rod and line if the fish are not there.

I talked to one old man who was born and bred in the

Highlands and has lived there all his life. He had worked on a big estate and, after a few generous drams, he told me how he had supplemented his modest income as a youngster with a spot of poaching. The local hotel owner did not ask too many questions when he was offered a fresh salmon at a knock-down price. Later the poacher turned gamekeeper and he worked as a ghillie, guarding a fine stretch of river and looking after well-heeled anglers from south of the border happy to spend a small fortune for the chance of a salmon on the end of their line. The old man pointed to a river in the distance.

'When the fish were coming back you could stand on the bank for hours and watch them fighting their way upstream,' he said. 'Good fish they were too and we never tired of watching them. Some years there would be more than others, but always there was enough for everyone. My father and grandfather caught them, and his grandfather too. I always thought they would be there for my grandchildren. It never occurred to us that one day they would stop coming, but now they have. It is a sad business.'

The figures support his story. The River Lochy in the Lochaber district was once the richest salmon river in the west Highlands. A member of the local fisheries trust showed me the results of a survey they had carried out to see what was happening to their disappearing salmon. Until the early 1980s catches of salmon were steady in all eight of their significant salmon rivers. In a good year anglers would haul in up to 1,200 salmon from the best beats on the Lochy. In 1998 they caught fewer than sixty. It is the same for the salmon's cousin, the sea trout. In rivers famous for their catches – the Shiel and the Moidart and the Ailort and many more – the graphs show an identical picture. In the eighties the catches are healthy. By the end of the nineties they have almost disappeared.

Apologists for the salmon farms say there are many other factors to explain all this, but they have great difficulty with two simple facts. Fact one: the decline of the wild salmon in the west Highlands coincides precisely with the growth of fish farms. Fact two: in other areas of Scotland where there are no fish farms catches have remained steady.

Why has it happened? Partly because wild salmon catch the lice and diseases of escaped farmed salmon, but also because fish farms are destroying the waters that give them life. I had read endless accounts of the pollution created by the salmon cages, heard endless tales of the filth they generate, but I wanted to see it for myself. That meant prowling about on the bottom of the sea – and that meant learning to dive.

My only previous experience of diving had been in the swimming pool of a friend in what was then called Rhodesia. I spent many boring days in Rhodesia when the guerrilla war was coming to an end and the politicians had begun to sit around the negotiating table. The life of a foreign correspondent can be dramatic, exciting and – in the tribal trust lands at the height of a guerrilla war – more than a little dangerous. It can also be deeply, deeply dull, and never more so than when you are waiting for meetings to reach a conclusion. It was on just such a boring day that my friend suggested we should all take a few hours off and go for a swim, leaving one volunteer behind to phone us when the meeting was about to break up. The added attraction – not that we needed one – was that he had just bought a set of diving equipment and we could all try it out.

So we took it in turns to sit at the bottom of his pool, cylinder strapped to back, cooling off from the baking heat of an African summer. Correction: some of us took it in turns. One of our number – a particularly competitive and aggressive young hack – refused to surface when his time was up. It was most annoying, but there's not much you can do when someone sits on the bottom of the deep end and refuses to budge. He lived to regret it. While he was down there, hogging the diving equipment and oblivious to the rest of us, the call came through. The meeting had ended and it looked as if there was a good story out of it. So we all cleared off and raced back into town and left him at the bottom of the pool. To this day I wish I could have seen his face when he surfaced to find that we had all gone – and when his foreign editor wanted to know how he'd managed to miss the story. 'Sorry, boss, I was sitting on the bottom of a pool at the time and the other buggers abandoned me.' Serves him right.

Diving in a swimming pool is one thing. Doing it in a Scottish sea loch is quite another. You could reasonably argue that my training was on the limited side, so when I tried to persuade a friend and professional diver, Simon Adey-Davies, to help me out I had to lie a little.

'Any experience?' he said.

'Oh sure,' I told him, 'but I'm probably a little rusty; it's been a few years now.' More than twenty to be precise and even then less than five minutes at a depth of approximately five feet, but I reckoned that if I went into too much detail he might not be too keen to take me. He looked doubtful but agreed – just so long as I did everything he told me to do. No problem there. I was terrified.

Our first dive was off the beach of a glorious bay in the Sound of Mull. I wanted to see how the sea bed ought to look when it is unpolluted so that I could make comparisons. Diving is not an elegant business. Simply getting into a dry suit is a serious performance. You have to peel it on and it must be tight: so tight you need to grease the wrists and the neck to get your hands and head through. The belt and the cylinder weigh so much you feel that the only way you'll be able to get down to the water's edge is if they fit wheels to you. Instead Simon stuck two huge fins on my feet and that meant having to walk, or stagger, backwards. But it was when he stuffed the regulator into my mouth that I had serious second thoughts about the whole exercise. Snorkelling is fine; you breathe through a tube but you can always spit it out and stick your mouth above the water. You can't do that when you're weighed down on the sea bed and about as mobile as an underwater pig in a corset. But I managed to stop myself panicking and it was worth it.

The water was clean and pure and it was extraordinary to share it with the fish and the crabs. It may be entirely unnatural to suck in air from a steel cylinder, and it's hellishly difficult to stop yourself rolling over like a drunken whale when the weight shifts, but you begin to feel at home in a foreign environment remarkably quickly. So, mission accomplished – part one of it at any rate. The next bit was much more difficult. I wanted to dive beneath a salmon cage.

I had been told what it would be like but I wanted to see it for myself. There was a problem. The owners of the big fish farms – multinationals now control more than seventy per cent of all the production – are extremely sensitive about their operations. That's hardly surprising. Most local politicians may have been reluctant to deal with the issue, but over the years many concerned environmentalists and local journalists have done their best. The resulting publicity has not stopped us buying the cheap fish but it has drawn attention to what is going on and there has been some modest tightening of the ludicrously lax controls. So the less of that sort of thing, from the point of view of the owners, the better. Hence the reluctance of Simon.

For a novice like me to dive safely beneath a cage would mean securing a line to the structure and there was not the smallest chance of being given permission to do that. So could we mount a clandestine operation? I asked a local diving expert. 'Sorry,' he said, 'I have to live here after you've gone.' There was, though, an alternative: a better one. We decided to dive to the bottom of a loch in the precise spot where a fish farm had been located one year ago. They'd had to move the cages because of disease and had left the waters to lie fallow.

One of the more persuasive arguments employed by the fish producers over the years has been that nature very quickly cleans up the mess created by man: the constantly shifting waters and strong tides in the bays and sea lochs means that it is never very long before the waters are back to their original pristine condition. By inspecting the bottom of the loch one year on we could put at least part of that claim to the test.

This time around I felt much more confident in my diving gear: eat your heart out, Jacques Cousteau. And, indeed, the first part of the dive went well. I was learning how to breathe reasonably smoothly and even how to keep my balance in the water. With all the foolish cockiness of an enthusiastic amateur I was starting to wonder what all the fuss was about. Plus there was the reassurance of knowing that Simon was always in sight. He had drilled me in three key signals that would tell him I was coping. Most importantly, if he made a circle of his thumb and index finger in an 'OK' sign, I had to make the sign back. If I did not he would assume I had a problem and he'd get me to the surface immediately.

That was fine in the clear waters of the bay and it was fine in the sea loch until we reached the area where the salmon cages had been and the water grew murkier. As we approached it was obvious that it was not only murky. This water and the sea bed was dead. There were no little fish darting around between the rocks and the vegetation. There was no vegetation. Instead the bottom of the loch was coated in what seemed to be a thick, black mud.

We swam down to the bottom. I had already discovered that one of the joys of diving with all the right equipment is that you can 'walk' along the floor of the sea on your hands, pulling yourself along from rock to rock or digging your fingers in the yielding sand. I had done it in the bay and it was great fun, disturbing all the little crabs and watching them scuttle off. I tried it here. Within a few seconds I was in a full blown panic. What had seemed like mud was not really mud at all. It felt like slime and the moment I dug my fingers into it to pull myself along, it rose in a great black cloud. God knows what this filth consisted of but I knew I wanted to get out of it. I desperately wanted to get to the surface and breathe clean air. Logic told me the air I was sucking in through the regulator was perfectly good, but logic has a nasty habit of deserting you when panic takes over.

I twisted and turned, trying desperately to remember the signal for 'Please get me up very quickly indeed!' and could not. Worse still, the cloud of filth was so thick I could not even see my instructor. Nor did I have the faintest idea when I was facing the sea bed and when I was facing the surface. I was terrified. Happily, the imperturbable Simon had a great deal more sense than me and, before we entered the water, he had floated a buoy in the loch and attached himself to it with a thin line. Then he attached himself to me, too. And even though I could not see him he knew the moment I started to panic. I don't suppose it took more than a few seconds to get me to the surface and I was never, in truth, in the slightest danger, but it seemed a great deal longer. When I was bobbing safely again on the top I pulled the regulator from my mouth, made some lame excuse about having to surface because of water getting into my mask, and said please could we go home and get a drink of something hot and strong. The next time I go diving I shall try somewhere a little more congenial.

But I had seen what I had come to see. The fish farm operators may try to argue that these conditions were exceptional, that if I had chosen an area where there was a stronger tide the muck would have been dispersed. They may say that if I go back next year or the year after that, the water will be clean once again and the life will have returned to the sea bed. Well, any or all of that may be true but my own direct experience, limited as it was, suggests otherwise. My diving instructor has swum and dived in those waters for years and he does not believe it either.

In any case, the facts speak for themselves. Imagine a sizeable coastal town pouring its untreated sewage into the sea and the effect that will have on the water. It has been estimated by the Swedish Royal Academy of Sciences that a fish farm producing one hundred tons of salmon every year spews out the equivalent waste to that of a small town. If you measure the sewage from all the farms in Scotland it adds up to the equivalent of a dozen sizeable cities. And this is raw sewage including phosphorous, nitrogen and ammonia going straight into the coastal waters. Nine thousand tons of ammonia waste is poured into the waters off the west coast of Scotland alone. Its effect can be devastating.

Allan Berry is a marine toxicologist who used to farm shellfish in those waters. He believes sea cages should be shut down. He told the *Sunday Herald*: 'Discharging large volumes of untreated slurry from sea cage fish farms into sheltered coastal waters is probably the best way, outside of the laboratory, to promote the growth of harmful algae and the production of toxins.'

French scientists have found that the ammonia in salmon waste stimulates the growth of a toxin called alexandrium minutum. It is a particularly nasty toxin that causes paralytic shellfish poisoning which can also affect humans. In the closing decade of the twentieth century there were thirty-four separate bans on catching shellfish in these waters because of it. There is also evidence of the lethal effects on shellfish of the pesticides used by salmon farmers. The senior water official at the Scottish Executive, David Rogers, says laboratory experiments indicate that toxin production brought about by an imbalance of plankton and algae can be caused by azamethiphos and cypermethrin, two of the pesticides used regularly to kill sea lice. Another form of contamination that affects humans is amnesiac shellfish poisoning which can result in damage to the brain. Because of it, fishing for scallops was suspended in 1999 over 9,000 square miles off the west and north coasts of Scotland. Once again ammonia waste from salmon cages was believed to have played a part and the coincidence with areas of high densities of salmon farms was truly remarkable. The following year again virtually the whole of the western coast was closed to shellfish collection.

Not everyone accepts that pollution from salmon farming can be directly linked to the epidemic of shellfish disease in the 1990s. It may be pure coincidence that the epidemic occurred just as salmon farming was exploding. It may be that the muck on the bottom of so many Scottish lochs and inland waters is relatively harmless. Or it may be that some officials – desperate to protect the fish farming industry – require standards of proof that defy common sense. It may also be that by the time the proof has finally been established to the satisfaction of everyone it will be too late.

I spoke to the man who owns the land bordering the loch in which I dived. He painted a graphic and glorious picture of how it used to be before the fish farms arrived. Of how, for a few magical days at more or less the same times every year, the salmon would swim through the narrow neck of the loch from their epic journey in the Atlantic to begin the last stage of the journey to their spawning grounds. Every year, as the number of cages multiplied and the number of salmon grew smaller, he tried to tell himself that the next year would be better. He still tries to believe that, but in his heart he knows that a pattern has been established.

*

I have written about what fish farming does to the salmon, what it does to the marine environment and what it does to other species: shellfish and wild salmon. I believe we should also be concerned about what it does to another species: humans. After those moments of panic in the polluted loch I know that I personally have no desire to eat the flesh of the fish who have to live in that sort of filth. Even more worrying than the conditions in which they live is what has to be done to them to keep them alive and make them grow fast and fat. One way and another a typical farm salmon is exposed to dozens of different chemicals. There are chemicals to kill parasites such as the sea lice and chemicals to kill microbes. There are disinfectants, antibiotics and anaesthetics. There are antifouling agents containing copper oxide. There are powerful medicines and injectable vaccines to prevent the fish falling ill and there are chemicals to make them the right colour. It's worth taking a closer look at some of them. Consider, first, the colour of a salmon.

A wild salmon acquires its delicate pink colour as it forages the oceans. Its diet includes crustaceans, smaller fish, plankton and algae, many of which are rich in caratenoids called canthaxanthin and astaxanthin. But a caged salmon cannot forage and the diet it is fed leaves it a pale, insipid colour. That's not at all what a fishmonger in a supermarket wants to see on his shelves; everyone knows salmon is meant to be pink. The customer expects it. So the producers add synthetic canthaxanthin to the salmon's diet. It doesn't do quite as good a job as nature – the flesh becomes a much more vivid colour than the delicate pink of a wild salmon – but at least it's pink. So far so good. And synthetic canthaxanthin is safe – or so it is said.

Why then, you might wonder, is it banned by the government as a food additive? It is illegal to add canthaxanthin to food for direct human consumption. It can cause deposits of yellow particles on the human retina and it is believed that small children whose eyes are still developing are most vulnerable.

It took the British government more than seven years to ban the additive after advice from the Food Advisory Committee which said it should be banned immediately. But there are still no laws against adding it to animal feeds – chickens are fed the stuff to make their yolks look more yellow – nor to fish feed.

Perhaps, by the time you read these words, canthaxanthin will have been banned. If so, it will be one more chemical to add to a very long list. The procedure is all too familiar to anyone charting the progress, if that's the right word, of intensive food production. A new chemical is created to cure one of a hundred ills. It may or may not be granted a licence by the appropriate authority. Frequently objections are raised and overruled because there is no hard evidence to prove that the chemical is harmful. Years later the proof begins to emerge and, after more delays, the chemical may be banned. You have to wonder how much damage has been caused in the meantime. There is also the problem of illegal use after a ban.

One group of chemicals is used to kill parasites such as sea lice. What they have in common is that they are highly toxic; they have to be if they are to do their job properly. Over the years many have been declared illegal for one reason or another. In the 1980s dichlorvos was used to treat farmed salmon. In the nineties it was ivermectin. By the turn of the century it was cypermethrin. One chemical based on cypermethrin is Deosan Deosect. It is a potent endocrine disrupter, a 'gender bender' that interferes with the hormones. In the words of the Environment Agency it is 'around one hundred times more toxic to some elements of the aquatic environment than organophosphate dips'. Organophosphates are hugely dangerous substances. And yet there is clear evidence that some of these chemicals have been used illegally and, for all anyone knows, it is still happening, such are the pressures on the officials who are meant to monitor these things. As for the 'legal' chemicals, there were three medicinal products licensed for use in 1989. By 2000 that had gone up to twenty-six and another fourteen were under consideration. Colin Adams, a biologist at Glasgow University, says most of those are 'general purpose poisons for dealing with crustaceans'.

It was left to Friends of the Earth and the *Observer* to produce the evidence of illegal use. They managed to obtain sworn testimony from men who used to work on fish farms. Jackie MacKenzie said he had bought and used Deosan Deosect. The deadly chemical was mixed with sea water and sprayed into the salmon pens for over an hour. The 'treatment' had to be aborted at times because 'adverse reactions' were displayed by the fish. Another fish farm employee, Jonathon Davis, signed a statement for Friends of the Earth in which he said that over the course of three years 'we used cypermethrin so many times I lost track'.

He said the nets over the cages were raised to administer the chemical and went on: 'The treatment usually lasted for about an hour. We watched the swimming action of the fish and when we could see them starting to shake their heads, we stopped the treatment.' The firm who employed Mr MacKenzie and Mr Davis never denied their statements. At the time of writing, neither have they been prosecuted. Following those damning testimonies Scottish Quality Salmon were forced in July 2000 to throw the company in question out of their so-called Tartan Quality Mark. The following September traces of the banned toxin, ivermectin, were found in samples of farmed salmon. In some it was four times the legal limit.

At the end of their lives the fish are starved for ten days and withdrawn from the cages for the last thirty days. That is meant to deal with the problem of chemical residues. Yet there is growing evidence that the great cocktail of chemicals fed to them or used as pesticides may, even so, be harmful to human health. Some of the world's leading scientists are seriously worried.

In 1999 the World Health Organisation produced a technical report on the issues involved in aquaculture around the globe and its implications for human health. Three representatives from the United Kingdom helped prepare the report and it was funded by Britain's own International Development Department. It concluded that the 'gaps in knowledge hinder the process of risk assessment and the application of appropriate risk management strategies with respect to food safety assurance for products from aquaculture'. In other words, we don't know enough about what's going on to be sure that it's safe. What we do know, because the report says so, is that there is an 'urgent need to raise the awareness of fish farmers of food safety issues associated with farmed fish and of the impact of the consumption of contaminated food on human health'. It also called for studies to be conducted to determine 'whether the abuse of pesticides can result in residue levels in fish tissue that are potentially harmful to human health'.

You might think, in view of such concerns, that there would be some pretty tough monitoring going on. Not so. The body charged with the task is the government's Veterinary Medicines Directorate (VMD). In its annual report for 1999 it admitted it was having problems keeping a careful eye on farmed fish: 'Sampling is often more difficult than in other sectors because of the geographically hostile environment in which the fish are reared and also because treatment records are often not kept on the sampling site but at the company's main office in a different location.'

To that lame excuse I suspect most of us would say something like: 'Well get your act together and bloody-well make them keep their records where it is convenient for you and not for them. You are meant to be policing them, after all.'

The VMD also promised that 'sampling practice is improving with experience'. You can't help but wonder how much 'experience' they need. Fish farming has been going on now for a whole generation. Must we wait another generation? Another body, the Working Party on Pesticide Residues, reported in 1999 that one hundred samples of fish and fish products were analysed. Not one of them was from a fish farm.

But from the small amount of work that has been carried out with farmed fish it is clear there is cause for concern. Since 1995 the VMD has detected residues of ivermectin and chlordane in farmed salmon. Chlordane is an insecticide which has not been authorised for use in veterinary medicine or as an insecticide on growing crops in the UK for many years. In 1999 the VMD found illegal residues of oxytetracycline and one sample tested positive for aflatoxin. The VMD describes aflatoxin as 'toxic and carcinogenic affecting primarily the liver'.

There are fears that other toxic chemicals may be finding their way into farmed fish products in the feed they are given which may, itself, have been contaminated. In July of 1999 the *New Scientist* reported that toxaphene had been found in the blubber of seals. The journal *Chemosphere* reported high levels of the same chemical in fish oil and fish meal. It said: 'Results indicate that feed can contribute to toxaphene contamination of farmed salmon from Europe.' Toxaphene is another organochlorine insecticide. Like PCBs it is one of those poisons that stays with us for a very long time; it accumulates in body tissue and is passed up the food chain. Of all the harmful chemicals it is these, perhaps, of which we should be most scared.

*

One of the most dubious achievements of modern man is that we have managed to create an environment full of poisons which do not disappear like smoke from a chimney but hang around and may build up in our bodies. The scientists tend not to put it quite like that. They use expressions about lethal compounds like PCBs and dioxins such as 'very persistent' and 'ubiquitous in the environment'. And, because they are everywhere, scientists have to find a way of measuring them and defining what they call 'acceptable' levels in our bodies. Frequently they will return to work they have already done and revise previous recommendations. In 1999, for instance, the World Health Organisation sharply reduced the amount of PCBs in our bodies considered safe, yet the Department of Health in Britain refused to accept the lower figures.

As a layman, I can't help feeling that to call this sort of thing science is not much different from trying to cure cancer with the eye of a toad and the tail of a newt. And I know from talking to them that many scientists take the same view. What is an 'acceptable level' of a dangerous poison? Common sense demands that there are endless factors which will affect the level: age, general fitness, gender, exposure to other chemicals, predisposition to one of many illnesses, the history of one's parents, genetic make-up, the efficiency of one's metabolism . . . and on and on and on. And yet a report from the Ministry of Agriculture grandly informed us in 1997 that the 'dietary intake of PCBs and dioxins in the UK was shown to be comparable to other countries and does not give cause for concern'.

What, for a start, does 'comparable to other countries' have to do with it? If it were ten times as high but still 'comparable' would that be acceptable? Patently not. The sentence is a nonsense and so, many believe, is the science. Can any scientist put his hand on his heart and swear to us that by increasing our intake of PCBs and dioxins we are doing ourselves no harm so long as we are still within the recommended acceptable level? The question answers itself. The frightening truth is that we learn more every day about the deadly effect of these 'persistent' poisons and the message is always the same: we should have been more, not less, sceptical in the past.

But let us accept that there is, as the World Health Organisation would have us believe, a 'safe' level of dioxins. They are a by-product of many industrial processes, including the incineration of waste and smelting, and are amongst the most deadly substances known to man. Dioxins are implicated in causing cancers and brain abnormalities, in endocrine disruption and reproductive problems and much else besides. Their danger has been recognised for a very long time and, over the years, many controls have been introduced. Our exposure to them through food probably reached a peak some time in the seventies.

There is another group of chemicals that have very similar effects to dioxins: PCBs. It was not until 1997 that the Department of Health accepted the similarity. It also accepted that their toxicity can be measured and directly compared to the most poisonous of the dioxins. In technical jargon the toxic equivalent is called a TEQ. Once PCBs were included in TEQ calculations it became clear to scientists working in this field that many people were exceeding the 'safe' levels. This is where we return to fish, both wild fish and farmed salmon.

Pretty well all fish contain some levels of dioxins and PCBs. Oily fish, such as salmon, herrings and tuna, contain more because they have a higher fat content and the chemicals build up in the fat. We are told endlessly by health advisers that fish, particularly oily fish, is good for us. What is less well known – and was not widely publicised by politicians – is the content of a paper prepared for the Ministry of Agriculture by its Food Contaminants Division in 1999 on dioxins and PCBs. It found that if we eat fish more than twice a week we are likely to exceed dioxin and PCB intakes considered safe by the World Health Organisation.

The Ministry had carried out a study into our diets and consumption patterns in 1992. Seven years later it used that information to calculate our intake of dioxins and PCBs. The average adult would consume a TEQ that was slightly below the upper level recommended as safe by the World Health Organisation. But 'high level' consumers would be well above the limit. It was far worse for small children eating a lot of fish.

Miriam Jacobs, a toxicologist at the University of Surrey, has been investigating the presence of harmful chemicals in fish. She found disturbing levels of dioxins and PCBs in farmed salmon. In one fish she found that a single serving would put an adult's consumption of PCBs slightly over the limit recommended by the World Health Organisation. If a child ate that same fish she would be well over the limit.

The explanation seems to be simple enough. The caged fish are fed on small fish caught in the open seas and concentrated into pellets. Those small fish are, themselves, contaminated because of the way we have been dumping toxic waste into the seas over the years and the effect of the food manufacturing process is to magnify the concentration. Dr Jacobs accepts that her samples are too small to be relied on. What is truly remarkable is that the research that is needed to provide the definitive answers has not been carried out.

It is clear that any food production methods which increase the risk of additional chemicals entering the food chain must be looked at very carefully indeed. Fish farming is not going to go away. On the contrary, within a few years more than half the fish we eat will be reared in cages. The next big development is cod farming. Cod grown from eggs taken from wild fish caught off the west coast of Scotland have already been sold in British supermarkets. The next step is to rear them artificially from the start. By 2010 50,000 tons of cod could be farmed a year – almost a third of the total amount of cod we eat in Britain. Under farmed conditions it may be possible to produce a mature cod after only two years. In the wild it takes anything up to five years for a cod to mature. Once again, nature will need a little help from our chemical friends.

As the fish farming industry grows still larger so new forms of chemical and biological controls will be introduced. Many farmed salmon already have a diet of genetically modified soya and cereals. If the feed, why not the fish? That, too, has already happened. Scientists in the United States have created genetically modified salmon that grow up to six times faster than wild salmon. These monsters have not yet been farmed commercially, but many people believe it is only a matter of time before public opinion has been softened up and we have been persuaded that it really does make sense to introduce these genetic freaks into the food chain.

In the Brave New World of twenty-first-century aquaculture and in the desperate quest for a 'beefier' fish, scientists around the world have tried injecting fish with cattle, pig and even human genes. Field trials have already taken place in Canada, Hungary, New Zealand, China and Scotland. The company that developed the GM salmon says they are infertile and would not pose a threat to wild salmon. The Centre for Food Safety in Washington says it is simply not possible to give that guarantee.

We should also look carefully at some of the new vaccines being developed to keep farmed fish healthy and some of the new hormones to speed their growth. The Americans use a genetically modified hormone to get more milk from their cows; it has been banned by the European Commission because of links with cancer. Now a synthetic hormone has been developed that enables rainbow trout to grow seventy per cent faster.

The *New Scientist* reported in the summer of 1999 on a new 'cutting-edge' vaccine for farmed salmon and trout. The vaccination involves injecting DNA into fish. Most vaccines work because they contain material that triggers an immune response directly. When DNA is injected it is taken up by the cells, which then begin to produce the viral proteins and the immune system attacks the virus itself. The danger is that the DNA could be permanently incorporated into fish chromosomes. In other words, the fish might become genetically modified, with results that cannot be predicted. One of the scientists who developed the vaccine admits that that is a 'theoretical possibility' but the chances are tiny and, even if it happened, it 'should not cause problems'.

You may say: it's all very well to raise concerns but oily fish like salmon are good for us and that outweighs any small, theoretical risks. True, salmon is good for us. I can remember as a child my parents talking about fish as 'brain food'. They might not have known why, but the habit of one meal of fish a week was based on sound principles. For three centuries and more we have been taking cod liver oil in the belief that it helps with all manner of ailments, and so it does. It not only reduces the risk of heart disease, it also reduces hyperactivity in children; it's used for treating depression; it lowers aggression under stress and it can help people suffering the pain of arthritis. Oily fish such as salmon contain Vitamin A, which is good for our eyes; Vitamin D, good for bones and teeth; and some important trace elements, and there is yet more.

There are two families of essential fatty acids (EFAs) which our bodies need to grow and develop properly. They cannot be made by the body so we must get them from somewhere else. Salmon is high in omega-3 fatty acids. To be more precise, wild salmon is high in them. That is because of the food they eat as they forage and hunt in the ocean – smaller fish, seaweed, different forms of algae. The food they eat on a fish farm is different. Most of it will be fish meal, though that does not mean it will be entirely fish. Far from it. Some producers use poultry by-products and blood meal because it is cheaper and easily accessible. Some even throw in the chicken feathers. Once it's all ground up small you can't spot the difference. About seventy per cent of the feed will be fish oil and fish meal and the rest is soya, wheat, ash and other bits and pieces. According to some experts in nutrition, fish that are fed grain instead of pure fish meal will be relatively low in omega-3 fatty acids but abnormally high in omega-6 fatty acids. The difference between the two is that we usually have too little omega-3 in our bodies but we have too much omega-6. Feed is the biggest single cost in farming salmon, and the price of fish meal has grown rapidly over the past few decades. It will continue to grow as demand increases, so the pressure is on the farmers to use as little of it as they can manage. That may mean using more grain or more blood meal and bone meal.

I have written at some length about salmon farming because it seems to me a cautionary tale that exemplifies our cock-eyed approach to food over the past generation. The first thing we do is wreak enormous damage on the stocks of wild fish, first by over-fishing and then by bringing in laws which are meant to protect certain species but actually result in massive wastage because trawlermen throw back into the water vast quantities of fish that their quotas do not allow them to catch. The fish are, of course, dead by then anyway.

Then we say: 'Oh God! We're going to run out of fish. What can we do?'

And then we do precisely the wrong thing.

In the case I have described, we set up farms to rear salmon as though they are battery chickens. Because that kind of practice inevitably leads to disease we pour in the chemicals

– into the water and into the fish themselves. It is inevitable, too, that some cowboy producers will break the law and use poisons that have been banned, and which do immense harm. The pollution and the sea lice that thrive in intensive farming conditions then devastate the wild salmon, which were already under pressure for other reasons and which farm production was meant to help.

But the lunacy does not end there. Now all these farmed fish need to be fed. And if they are to survive they must eat other fish in one form or another. So, massive amounts of fish must be caught in other oceans of the world to keep the farms going. It takes two or three pounds of 'harvested' fish to produce one pound of farmed salmon. And this is meant to be reducing pressure on the fish stocks?

In the words of the highly respected *Nature* magazine: 'The use of wild fish to feed farmed fish places direct pressure on fisheries resources. But aquaculture can also diminish wild fisheries indirectly by habitat modification, collection of wild seedstock, food web interaction, introduction of exotic species and pathogens that harm wild fish populations, and nutrient pollution. The magnitude of such effects varies considerably among aquaculture systems, but it can be great.'

And the final offence is that when you go to your fishmongers or – more probably these days – your supermarket, what do you see? You see a wonderful array of fresh fish, much of it from foreign waters, and always plenty of salmon at a remarkably cheap price. Some supermarkets sell more salmon than any other fish. You may see it advertised as 'catch of the week', language clearly designed to make you think that only a day or so ago it was swimming powerfully through fresh, clean water. You may see little pictures of fishing boats or rod and line on packets of smoked salmon, which is equally deceptive. What you do not see is anything that even hints at the reality. All this salmon has come from a fish farm, most of it reared so intensively that when chickens are reared in a similar way many of us refuse to buy them or their eggs.

So what is to be done? Barring a disaster of BSE proportions there is no doubt that in the years to come there will be much more fish farming. In my view that is regrettable. It took mankind centuries to develop a system of husbandry for animals on land. It is arrogant beyond belief to *think* that we can learn to farm the seas in the space of a few years. Having begun the process of destroying one of the world's great natural resources – the fish in our oceans – we seem bent on accelerating the process. The cod was once so abundant it fed hundreds of millions of people in nations around the globe. In the summer of 2000 it was declared an endangered species.

If we are to farm fish off the coast of Scotland or anywhere else the least we can do is insist that we do so less intensively. And indeed less intensive fish farming is beginning to happen. Where the wild waters of the Atlantic meet the North Sea off the Northern Isles an organic fish farming industry is coming into existence. No toxic chemicals or artificial colouring agents are used and the stocking rates are about half the levels used further south. Because the waters of the bays flow so fast there is little danger that pollution will build up and other fish or shellfish will be affected. The fish meal is made from crushed shrimp shell and vegetables and the organic standards allow less than one quarter of it to be oil.

There are problems even with organic fish farming, though. Animal welfare is an issue for many people. They believe it is simply wrong to rear wild creatures in cages, whether they are fish or mammals, and we should stop completely the practice of farming carnivorous fish such as salmon, trout, cod and halibut. The medieval monks of Japan farmed carp because they live perfectly happily off vegetation. The farming of shellfish is also perfectly sustainable so long as they are not poisoned by the pollution from salmon cages.

Whatever approach we take, we must use fewer chemicals and use them sparingly. We must police the farms more aggressively and monitor their output efficiently. And if – in spite of everything – we still buy farmed salmon, we must demand to be told the truth about what has happened to it. It is not too late; aquaculture is still a relatively new industry. If consumers had been more demanding over the years we might have avoided one of the other catastrophes aided and abetted by intensive agriculture: the threat to antibiotics.

BATTLING WITH BUGS

Antibiotics

Sadly for the romantic view of history, many of the great scientific discoveries through the ages did not come about in quite the way our school history books tell us. Newton did not need an apple falling on his head to reveal the mystery of gravity and Archimedes did not run naked down the street shouting 'Eureka!' when he realised why his bath overflowed as he got into it. Nor did Alexander Fleming examine the mould growing in his Petri dish and rush around the place shouting: 'I've discovered penicillin! I've saved the world and I'm going to be famous for ever!' What he actually observed was that the mould was killing off bacteria and what he thought he had discovered, amongst other things, was a really useful antiseptic. In truth, he had opened the door to the age of antibiotics. They would eventually be capable of killing almost every bacterial infection known to man. As a result countless millions of lives have been saved. Take just one disease: tuberculosis.

Before antibiotics TB killed vast numbers of people around the world every year. In this country the death rate fell from its terrible heights in the nineteenth century as people were moved out of the slums, where they lived cheek by jowl in damp and dirty tenement blocks, into slightly more spacious homes. The slums were the perfect breeding ground for the TB bacteria. Even so, the disease persisted.

To be diagnosed with TB in this country meant being

packed off to a sanatorium. There, with hundreds of others, the patients would lie in bed gasping and wheezing. Sometimes they would get better. Often they would not. Even if they survived, their lungs would usually be horribly damaged, their lives left hanging by a thread. The next slight infection might turn to pneumonia and that would be that. The doctors could make a stab at dealing with the symptoms of that terrible disease; they could do nothing to cure it. That was down to chance. Either the body fought it off, or it did not. With the help of antibiotics it was no longer left to chance.

The real credit for the antibiotic era should go to two other scientists: Howard Florey and Ernst Chain. They believed that a whole range of antibiotics could be created to fight an enormous variety of diseases which had no cure. And they had a hunch that Fleming's penicillin had potential as an antibiotic. What they did not have was money for research. So they went to the Medical Council. The idea of producing an antibiotic from a mould did not go down terribly well but in 1939 they were given a grant – for the magnificent sum of £25. Even in 1939 you couldn't exactly change the world with £25. So they did what so many scientists have done over the years – they went off to the United States where the Rockefeller Foundation gave them £9,000.

They came back to Britain and set up a research team in Oxford. They tried to patent their work but, once again, were rebuffed – this time on the basis that 'medical discoveries are for the good of mankind'. So, yet again, they turned to the Americans. Again, they were welcomed with open arms. The Americans succeeded in registering international patents for the new antibiotics and for all the research methods and findings of Florey and Chain. Thus was American domination of the world's antibiotics industry founded. It predominates to this day. Income from penicillin sales and royalties paid for a massive search for new antibiotics and the development of dozens of other drugs. Some of the companies that had to pay the royalties were British. The country that had discovered penicillin was reduced to paying the Americans for the privilege of using the knowledge.

In the early 1950s there was another accidental discovery involving antibiotics. This, too, was hailed at the time as a breakthrough, though for different reasons. It happened not in a laboratory, but on a chicken farm in the United States.

The newborn chicks were being fed on a mixture of food that included the waste mash from the production of an antibiotic known as aureomycin. It was assumed that the feed might have some beneficial effects, perhaps reduce the likelihood of the chickens being infected by various bugs. But something else happened as well. The chicks began to grow at an extraordinary rate. As the *Daily Telegraph* reported in its headline at the time: 'Drug speeds 50% growth effect on animals'.

A new industry had been born: antibiotics as growth promoters. Countries like Britain and America were already taking the first tentative steps towards factory farming. This made it possible in a way no one had previously envisaged. Yet there are many people who rue the day it happened because it threatens to undermine the great achievements of men like Fleming and Florey and Chain. Their work had triumphed over some of the most devastating diseases in the history of civilisation. This 'discovery' was not only to undo much of that work but, over the years, to create new threats to our health.

Ever since antibiotics became widely available doctors and scientists have worried about what might happen if the bugs they were meant to destroy learned to outsmart them. Bacteria have always been able to mutate, just as every other life form evolves; they just do it more quickly. It was always known that drugs which set out to attack them would force them to adapt. Under the right conditions many of them might be able to develop a resistance to a particular antibiotic. That might not matter too much if there were another drug to switch to. But what if there were no alternative? What if, every time a new and effective drug was developed against a dangerous microbe, the bug found a way of adapting and staying one step ahead? What if, each time the scientists came up with a new form of attack, the bugs developed a new form of defence? How soon would it be before we were faced with an army of superbugs that fought off every assault the researchers were capable of launching? Might we then be in an even worse state than we were before Alexander Fleming watched his mould grow?

It's the stuff of science fiction: hideous, giant bugs with their terrible fangs bared as they advance on the good guys, bullets bouncing off their invincible hides. But, as every doctor and every hospital nurse knows, it's happening. The fact that in real life the bugs are so small they can be seen only under a powerful microscope makes them no less dangerous. They are as lethal as any fang. Nor is this a threat that has exploded without warning. Those brilliant pioneers, Florey and Chain, had said back in the thirties that resistance to penicillin would become a problem sooner or later if we did not treat it with respect. The warning was ignored. By the time penicillin was in widespread use in the next decade there were signs that some strains of E. coli and other bacteria might be starting to develop resistance. The speed of that development has been startling.

One of the most common bacteria is called staphylococcus aureus. It lives in the intestines of some animals and is also perfectly at home on our skin. Mostly it is no problem but some strains are dangerous. They are a frequent cause of infection in hospitals. They infect wounds, can cause pneumonia and are highly contagious. In the worst circumstances they can kill. In the 1940s about ninety-five per cent of staph. aureus were sensitive to penicillin. By the 1990s ninety-five per cent were resistant.

Other forms of potentially deadly bacteria that had once been relatively easy to kill are now able to fight back against the drugs. Diseases that had been defeated are making a return. In hospitals an increasing number of people are dying, not from the disease for which they were being treated, but from infections picked up during treatment. Hospitals have always been dangerous places but in the early years of antibiotics hospital-acquired infections could usually be seen off. By the 1990s it was estimated that as many as 15,000 patients were dying from infections against which no antibiotic was effective. In intensive care units, where the lives of so many patients hang in the balance, there is one chance in two of picking up an infection. This is not science fiction; it is fact. The superbugs are with us.

Some are related to those staph. aureus. They began by fighting off humble penicillin and then fought their way through the ranks of the ever more powerful drugs that followed: erythromycin, streptomycin and many others. They saw them all off. They are now known as MRSA (methicillin-resistant staphylococcus aureus). Another group of superbugs that have followed in their victorious wake are known as VRE (vancomycin-resistant enterococci). It is these bugs that together are thought to be responsible for many of those 15,000 deaths.

They are also causing mayhem in the running of many hospitals. One of the biggest in the country – the Portsmouth Hospitals NHS Trust – was forced to shut down most of its operating theatres in the summer of 2000 because of them. I spoke to one orthopaedic surgeon who told me he had not been able to carry out a single operation in the Queen Alexandra

Hospital because they could no longer sterilise their equipment properly. For four months orthopaedic surgeons did not carry out a single non-emergency operation. It was simply too dangerous. Ten surgeons wrote to the hospital's chief executive to say that emergency bone operations had to be halted within a fortnight because of the problems with sterilisation. They said: 'We now feel that people's lives and well-being are at serious risk.'

In 1999–2000 no fewer than 480 patients contracted MRSA in the two Portsmouth Trust hospitals – twice as many as in the year before. One person died. She was twenty-two years old. By the autumn of 2000 the problem had become so serious that a nearby private hospital refused to admit patients from the NHS hospital unless they provided two skin swab tests three days apart to prove that they were clear of MRSA. The hospital denied that there was a link between the state of their sterilisation and disinfection unit and the level of MRSA. They said patients were catching MRSA 'in the community'.

Whatever the reason the bug has become a serious problem. In some ways what is even more worrying is what might happen to some of the most common bugs, the sort that are not confined to hospitals and live in our nose or throat or gut. If they succeed in getting into our bloodstream, perhaps through an open wound, and set up an infection they are fairly easily despatched with the right drugs. At least, they have been until now. But the older forms of antibiotics have been proving less and less effective and some of them are now useless against a whole range of bugs. Over the past decade the alarm bells have been ringing louder and louder.

In the late nineties a research programme was launched to try to find out what exactly is going on in hospitals around the world. It was led from Britain by Dr Bob Masterton of the Western General Hospital in Edinburgh. Over a period of three years Dr Masterton studied nearly 5,000 samples of common bugs taken from patients in ten hospitals: some young, some old, some sick and some recovering. The overall trend of the survey is that resistance to the older antibiotics is increasing. Resistance to meropenem, a relatively new and powerful antibiotic, remained at a very low level and did not seem to be increasing year on year. The research programme was funded by the drugs company AstraZeneca, which manufactures meropenem. But even that drug is powerless against the two groups of superbugs. Neither MRSA nor VRE were included in the study.

When I spoke to Dr Masterton, soon after he had published his interim findings, he was not exactly cracking open the champagne. He told me it would be foolish to assume there was no longer anything to worry about: 'There are three groups of bacteria that have developed resistance and they are a serious worry. As for the rest, there's no way of knowing how they will behave over the next few years. We can't possibly say that they will remain dormant. It is vital that we keep monitoring them to see how they are changing. We have only one real line of defence left against them now and we don't know of any new antibiotics being developed.'

AstraZeneca themselves said: 'The need for new antibiotics may never have been greater.'

Three-quarters of a century after penicillin was invented, we are on the brink of what could prove to be a medical catastrophe. Nor can we comfort ourselves by pretending that it won't happen to us.

We could certainly reduce the risk by making sure that we never end up in hospital. We could become vegetarian and thus avoid eating meat contaminated with resistant bacteria. Then we would also have to give up eggs and milk, both of which are a common source of infection. Fruit and vegetables could be tricky because some plants can carry resistant bacteria and transfer them to people. In some countries from which we buy food it is perfectly legal to spray some crops with antibiotics against bacterial disease. Having done all that we would have to avoid contact with any other humans who might be carrying resistant bacteria. It could prove a lonely, hungry existence.

How much of this is a result of what has been happening on the nation's farms and in its broiler sheds and battery farms and pig units is difficult to estimate. A report to the government in 1999 by the advisory committee on the microbial safety of food said the growth of superbugs could be blamed directly on the use of antibiotics on farm animals. It said up to half the use of antibiotics was in agriculture and nearly every antibiotic used on humans has been used to treat farm animals.

There is no doubt that patients and doctors are also to blame. Too many of us get a sore throat and rush off to our GP, croaking miserably: 'I need an antibiotic.' All too often the doctor takes the path of least resistance. He knows perfectly well that antibiotics are useless if we have been attacked by a virus, but he's likely to be tired and overworked and has another dozen patients to see before lunch. From bitter experience he knows we shall make a fuss if he does not hand over a prescription and he also knows that if there's even the most remote possibility that he might make a mistake by refusing to prescribe an antibiotic he may end up being sued for negligence. So, as often as not, he gives in and we toddle off to the chemist clutching our prescription.

Then we take the medicine for two or three days and our throat gets better – as it probably would have without the medicine – and we throw away the rest of the pills before we have finished the course. That's pretty well perfect from the perspective of a harmful bug. There is just enough of the drug in our system to help it develop resistance so that when we really need that particular antibiotic to kill it, it will have no trouble surviving and multiplying. But the drug may also have succeeded in killing off the harmless bacteria that live side by side with their dangerous cousins. That means the nasty bugs can move into their space and they have yet more opportunity to multiply. So, one way and another, when we most need a specific drug to kill a specific type of bacteria, it may no longer be worth taking.

A curious effect of this over-prescribing has been seen in the United States, the country where doctors need a law degree to protect them against their patients and where the only vital document in their desk drawers is the insurance policy. There, children with serious ear damage from repeated infections are far more likely to come from rich neighbourhoods than poor. That's because medicine is expensive in America. When a rich mom takes her toddler to the doctor with a slight ear infection she'll come away with an antibiotic. By the time the toddler is in high school he may have had half a dozen courses of antibiotics and his bugs are resistant. Hence the damage. The poor mom, by contrast, cannot afford the medicine and in this respect at least her baby benefits.

There are at least as many antibiotics used on the farms and in the broiler sheds of Britain as there are in all the doctors' surgeries and hospitals put together, which takes me back to the discovery they made on the chicken farm in the United States in the fifties about antibiotics as growth promoters.

The British Poultry Meat Federation does not approve of the term 'growth promoters'. It says the bird does not grow any bigger than it otherwise would, so the phrase is misleading. The drugs are merely 'digestive enhancers'. They free more nutrients for absorption by the bird, thereby improving the efficiency of feed conversion. To which I would say: 'So what? The effect is the same. You feed the bird antibiotics and it grows more quickly.' The history of drug use on

Britain's farms proves, yet again, how careless we have been with embracing new technology without fully considering all the possible consequences.

In 1953 Parliament passed the Therapeutic Substances (Prevention of Misuse) Bill. The name turned out to be somewhat misleading. Whatever else it may have prevented, it was not misuse of drugs. Until then the government had restricted the use of penicillin to prescriptions only, both medical and veterinary. What this law did was relax the regulations to allow small quantities of penicillin and tetracycline to be included in animal and poultry feeds to help them grow more quickly. They did not need a prescription from a vet. The bill had an easy passage through the House of Commons.

A few of the more concerned MPs enquired gently about such possible risks as drug residues ending up in our meat and some of the more dangerous bugs acquiring resistance, but any doubts they may have had were easily set to rest. Not only the Minister of Agriculture but the Minister of Health, Iain Macleod, urged them to pass the bill. Mr Macleod told MPs: 'I am assured by the Medical Research Council [. . .] that there will be no adverse effect whatever upon human beings.'

Only one MP spoke out against the bill with any passion. He was Colonel Gomme-Duncan, one of the many hyphenated army officers in the House in those days. 'May I ask whether we have all gone mad to give penicillin to pigs to fatten them?' he asked. 'Why not give them good food, as God meant them to have?' His colleagues shook their heads sadly. The silly old buffer just didn't recognise progress when he saw it. He looks rather less of a silly old buffer today.

Yet, given the climate of the times, it is easy to see why politicians rushed to embrace this new approach. The war had been over for only a few years. Most food was still rationed. The idea of bacon and eggs sizzling in the pan for breakfast every morning and a chicken roasting in the oven for Sunday lunch was the stuff of glorious dreams for most families who had come through the war on dried egg powder and tinned Spam. Anything that could feed the nation more efficiently, nutritiously and appetisingly was to be welcomed and it seemed this was the answer. The Agricultural Research Council had conducted trials in which about three-quarters of all the pigs and poultry given antibiotics on a daily basis showed increased growth rates. There was another factor too: Britain was importing a large amount of grain and other livestock feed. If feeding animals on drugs meant they would need less imported grain, the nation would benefit economically. There really was no argument.

But worries about resistance to antibiotics were beginning to niggle away. Even as MPs were voting on the new Bill doctors were worrying that so many bacteria were showing resistance to penicillin. They were told not to worry; it was inevitable that some bacteria would mutate but there would always be new forms of drugs to deal with them. In 1959, a Japanese scientist, T. S. Watanabe, had claimed that antibiotic resistance could actually be infectious. In other words, the resistance could be transferred from one type of bacteria to another inside the alimentary system of humans and animals. This was a different matter altogether. If it was true it meant that Macleod and the Medical Research Council had been wrong. Feeding animals low doses of antibiotics on a daily basis really could pose a threat to human health. The government responded by doing what politicians always do when they are faced with something they do not understand: they set up a committee, the Netherthorpe Committee, in 1960. What a mistake that turned out to be.

Not only did Netherthorpe conclude that feeding antibiotics to animals was quite safe, it recommended that the use of growth-promoting drugs could be extended to calves under the age of three months. It was the equivalent of sending an invitation for a win-a-free-holiday contest to every potentially dangerous bug in the land: 'Enjoy a fabulous free fortnight and you'll leave feeling fitter than ever. You'll be a new bug!' The invitation was enthusiastically accepted.

During the sixties things began to go badly wrong. There was a rash of serious outbreaks of salmonella food poisoning that proved resistant to a range of different drugs. It seemed that Watanabe's claims were being vindicated. Not only that, but farming was changing in Britain in a way that meant serious food poisoning was going to become a real threat to the nation's health.

There is a curious paradox here. Antibiotics killed the bugs that gave us some of the most deadly infections. And yet we were about to expose ourselves to a system of farming that greatly increased our risk of food poisoning. The explanation lies in the way antibiotics enabled some farmers to change the way they reared their animals. Not only would the animals grow far more quickly; they could also be reared far more intensively.

Left to their own devices, creatures such as pigs and chickens like to root around and peck in the dirt outside. It's what nature intended them to do – hence the tough snout and the sharp beak – and it's good for them. They dig out many of the vitamins and minerals they need to stay healthy. They also come into contact with a wide range of bacteria and that helps them develop strong immune systems. One old pig farmer told me some years ago that pigs need three things to thrive: food, freedom and a dry bed. But they use a great deal of energy rooting around. If you want an animal to grow fat and to do it quickly, you stop it moving around and keep stuffing it full of feed. It's the difference between an active youngster who rushes about all day and another who flops on a couch and watches the box, endlessly snacking on crisps and chocolate bars.

But when farmers first began trying to turn animals into couch potatoes by cramming them into poorly ventilated buildings with food on tap they created more problems than they solved. If one chicken fell ill, the disease could spread through the rest of the flock before the farmer realised what was going on and had a chance to treat the sick birds. Out in the open a sick chicken is a sick chicken. In battery or broiler shed conditions it's a financial disaster. So antibiotics made factory farming both possible and profitable.

Not even the most irresponsible doctor could imagine doing what many farmers have been doing routinely over the years. It amounts to sprinkling a dose of antibiotics on our porridge in the morning in the hope that it will stop us getting ill in the first place. Eighty per cent of antibiotics were being used in agriculture not because the chicken or the animal had fallen ill but because it might fall ill or because the farmer wanted it to put on weight more quickly. It made financial sense for the farmer – and for the shopper.

When I was a child chicken was considered a great luxury. We had it on Easter Sunday as a special treat. I remember to this day how we kids fought over who got to pull the wishbone. I can't somehow imagine modern children doing that – assuming they even know what the wishbone is. We eat so much chicken now that there would be a wishbone for every child in Britain to make a wish every other day. About 800 million broiler chickens are raised in Britain every year. As everyone who has ever been anywhere near a supermarket knows, chicken is cheap. That's hardly surprising when you look at how they are reared.

A broiler chicken is big enough to be slaughtered only forty-one days after it has pecked its way out of its mother's egg. That is half the time it took forty years ago. For the first few days of its life, when it looks like something off an Easter card, it weighs less than two ounces. If it were living with its mother in normal circumstances, ranging freely, it would shelter under her wing during its early life. She would protect it and make sure it was fed. In a broiler house it has no protection and must fend for itself.

There have been some improvements over the past few years in the way chickens are reared in this country, but the sheds in which they lead their short lives are still not for the squeamish. I cannot pretend that what follows is based on my own observations – though that's not for want of trying. Over the years I have tried many times to get permission from the owners to film in broiler sheds. It has always been refused. I have, though, seen film shot covertly with a hidden camera and spoken to reporters who have worked in the sheds. They themselves had to lie to get in: to pretend that they were casual workers wanting to earn a few pounds. Some of the owners have denied that things are as bad as they describe and insist that the chickens are reared as humanely as possible. So why, you might wonder, do almost all of them refuse to open their doors to the occasional journalist? I leave it to you to decide.

There are usually about 25,000 birds in each shed. The building has no windows, only artificial light. Until relatively recently the lights were kept on for all but thirty minutes out of twenty-four hours. Chicks grow more quickly when they are in the light. After some aggressive campaigning from animal welfare organisations and in spite of squeals of protest from the chicken industry, that was changed. They now have eight hours of light followed by eight hours of dark. In almost every other respect their lives are hellish.

The weakest birds die quickly. They simply cannot compete for food. Those that survive put on weight at an extraordinary rate. The fatter they get the less space they have. By the time they are approaching their killing weight they will each have the space they can stand up in – just. If you opened this book and laid it flat on the floor, that's about it. Imagine a fully grown chicken surviving in that space. Many do not. The death rate is usually at its highest on the hottest days when the ventilation struggles to cope. Then many of them suffocate to death. It is estimated that more than forty million young birds die in broiler sheds in Britain every year. That is about seven per cent of the total. In the worst of the broiler sheds the dead are left for lengthy periods where they fall, their bodies rotting.

At its peak, a broiler chicken will be growing twice as fast as a hen reared to lay eggs. Its diet is not designed to produce a healthy body and healthy bones. It is meant to produce as much flesh – especially breast meat – as possible. Consequently its bones cannot cope with all the weight. Animal rights campaigners claim that nine out of ten birds are crippled from early on. Their bones almost literally crumble. Some of them have almost to crawl to the feeders – a bit like a baby shuffling on its bottom, though the chickens use their thighs to move them along. There is no certain way to prove that they are in constant pain from broken bones or crippled joints, but it has been observed that when they are given a choice of two feeds they will choose the one that contains a painkiller.

The birds also suffer from the diseases that would afflict any animal that is grossly overweight, particularly heart and lung failure. Some diseases are unique to their hideous environment, such as blisters that develop on their breasts from being constantly forced up against other birds. At the end of their days they are taken off to be slaughtered. A merciful release, you might think, though there is little enough mercy shown in the killing process. Everything must be done at the maximum speed possible. Time, after all, is money. There is only a small profit made from a single bird; this is a volume business. So they are grabbed by their legs – two in each hand – and crammed into crates. At the slaughterhouse the chickens are hung, upside down, in shackles on a conveyor belt. They are then dragged along and their heads are dunked into a bath of water that has a powerful electric current running through it. The theory is that the bird is instantly stunned to save it any more suffering. The reality, according to some who have witnessed this gruesome spectacle, is that many of the birds jerk their heads up and are still conscious when their throats are slit. The whole process – from the egg cracking open to the chicken appearing on the dinner table – takes about six weeks.

The reason I have written about this process at some length is not to shock with the detail. I happen to take the view that this way of rearing chickens is cruel and unacceptable, but others disagree. They argue that it is foolish to have such scruples when it comes to poultry farming. They say it's simply not true that broiler chickens suffer and that they are perfectly content so long as they are warm and dry and well fed; if they were not they would not thrive. Others will confess that they hate the idea of broiler sheds but are reluctantly prepared to tolerate them if they result in cheap food. Better, they argue, that poor families should be able to afford chicken for the children than that the chicken should enjoy a good life.

But this book is about the safety of food and it is in that context that we should be concerned about what happens in a broiler shed. Since the intensive rearing of poultry and battery hens began there has been a massive increase in the amount of food poisoning from eggs and chickens. In a typical year there are probably about one million cases across the country. Many of them are serious. Some of them are fatal. Edwina Currie had her fifteen minutes of fame – or notoriety, if you prefer – when she was a health minister in the late eighties. She famously told a television reporter that pretty well all the eggs on sale in our supermarkets contained salmonella. That statement caused enormous damage to the egg industry. Mrs Currie was savagely attacked by every other egg producer in the land and soon got the sack from the government. But she was only telling the truth.

The number of cases of salmonella poisoning reported to the Communicable Disease Surveillance Centre rose dramatically after factory farming began. In only ten years – from 1980 to 1990 – it trebled to more than 30,000 and those are just the reported cases. In the closing years of the last century, there were sharp falls. One reason for that may be improved hygiene on battery farms, but there has not been a similar drop in other types of food poisoning. The real reason may prove what the critics of antibiotic growth promoters have been claiming right from the beginning: the drugs actually increase the variety of dangerous bugs in a chicken. That may seem a perverse conclusion. You would expect the drugs to kill bugs rather than encourage them to grow. In fact, it makes perfect sense.

The insides of an animal – or a human for that matter – are home to a vast population of bacteria. There are trillions of them and they are capable of multiplying at a phenomenal rate. Given the right conditions E. coli bugs can double in number every twenty minutes. In twenty-four hours you would have more than two billion bugs. Not for nothing are they described as the rabbits of the bacterial world. In spite of their reputation, most E. coli are pretty harmless creatures, just like most other bacteria. But they have a habit of mutating all the time: one mutation for every billion divisions. So it can produce a very large number of mutants in a very short time. Again that need be no great problem. The mutants might actually weaken the bacterial strain. But if you then introduce them to a steady dose of antibiotic of one sort or another there is no way of knowing what may happen. It may be the microbial equivalent of feeding spinach to Popeye, allowing one mutant to become predominant.

There is another factor too. Most bugs are in competition with each other for space in which to live and nutrients on which to feed. There is a constant battle between the good guys and the bad guys and if we disturb the environment too much we can give the bad guys a chance to squeeze out the friendly ones. All bacteria fall into one of two categories: gram-negative and gram-positive. The bugs that give us food poisoning are gram-negative. The antibiotics that help poultry grow faster are gram-positive. The result is that the drugs kill the bacteria with which salmonella bugs would normally be in competition.

The more salmonella in the chicken, the more likelihood there is that we will fall ill when we eat it. There is a myth that salmonella is relatively harmless. That's true for about ninety-five per cent of us. To the other five per cent who may have no resistance for one reason or another it can be highly dangerous. Over the past fifteen years or so a new and particularly nasty strain of salmonella known as typhimurium 104 has taken hold. Its resistance to the antibiotics which once killed it has increased from five per cent to ninety-five per cent.

Chicken farmers are well aware of the dangers from salmonella. Some of them inject their chickens with bacteria to try to restore the balance. So we have an extraordinary vicious circle. Chickens are not only fed antibiotics to make them grow faster; they are also treated with different ones to try to keep them fit. If they fall ill, they are treated again to kill the infection. But that kills beneficial micro-organisms. And so it goes.

In Sweden they started to ban the use of growth-promoting antibiotics in 1984 because they were so worried about the increasing amount of salmonella in their poultry flocks. Within two years they had banned them all. There is now very little salmonella in chickens produced in Sweden.

Another dangerous bug that can infest chicken farms is campylobacter. There is reliable evidence to suggest that nine out of ten chickens on sale in the supermarkets of Britain are contaminated with it. That statistic is based on research by government agencies which bought fresh chickens and tested them. Nothing very complicated there, but it is far more difficult to reach any detailed conclusions on what effect the bug is having on the nation's health. Once again, the amount of information is limited. Once again, the vital research work has simply not been carried out. But let us look at what we do know and try to draw some conclusions.

If we are scrupulously careful in the way we prepare and cook our food there should be no problems. The experts tell us that chicken should always be cooked thoroughly and we should take great care when we handle it in its raw state. Sadly, we are not always scrupulously careful and, unless human nature is about to undergo some dramatic transformation, many of us never will be. In the early nineties about 38,000 cases of campylobacter were reported to the Centre. Within six years the figure had passed 58,000. But those statistics reveal only a small part of the story. Research into surveys in the United States shows that for every case reported to the public health authorities there are at least another ten that go unreported. There is no reason to believe it is any different in this country.

Most cases go unreported because the hapless victims suffer nothing more unpleasant than two or three days of bloody diarrhoea. They may be treated by their GP, who may not bother to send off samples for laboratory analysis, or they may simply not bother going to the doctor at all. In that case, you may conclude, perhaps this is not a particularly serious illness. It is. Those who get away with a few days of discomfort can count themselves lucky.

Campylobacter poisoning can lead to severe septicaemia,

an infection of the blood which means at best a spell in intensive care and, at worst, serious health problems for a long time afterwards. It can even lead to something called Guillain-Barré syndrome, a neurological disease that causes paralysis and can be fatal. It is estimated that ten per cent of people struck down with the campylobacter bug will suffer some form of complication. But even without severe complications, this type of poisoning is serious.

So why? Why should there be so many cases and why do so many lead to such terrible complications when we know the bug that's causing all this? For both those answers we have to turn once again to the way poultry farmers have been using antibiotics.

Professor Hugh Pennington is perhaps the leading authority in Britain on medical microbiology. It was to him that the government turned when a large group of mostly elderly people were struck down with food poisoning in Scotland in 1996. They were all the victims of E. coli 0157. Twenty-one people died. Professor Pennington was called in to set up an inquiry and investigate. His credentials, knowledge and experience are not in question. He also has the ability – relatively rare in the academic world – to speak in a language that a layman can understand. He believes that while the reasons we are facing such a serious problem are relatively simple, the answer to the problem is not.

Rearing many thousands of chickens by cramming them together in one shed is the perfect way of encouraging the spread of disease. Professor Pennington draws an analogy with people who lived in the worst kind of slum in Victorian times. If one child came down with an infectious disease there was a very strong chance that it would spread through the whole community. Bugs thrive best where there is most overcrowding. Antibiotics make overcrowding possible. If they are used routinely they may not kill every nasty bug in every chicken, but the infected chicken will survive to spread the infection. In the old days, when chickens pecked around in fields and yards, some would die but the disease would probably not spread through the entire flock.

Nor, in the old days before factory farming, were chickens slaughtered in the way I have described. Professor Pennington has been around a long time in the public health arena, but he admits to being shocked at the conditions in conveyor-belt slaughterhouses.

'Even if they are not already contaminated when they go in alive, it's a reasonable bet that they will be by the time they've been slaughtered and are ready to be eaten,' he told me.

All of that may explain why there are so many cases of campylobacter. The reason why it is becoming so difficult to treat is closely linked.

One of the most important antibiotics used to kill the campylobacter bug over the years has been ciproflaxacin. It is one of the most modern of the fluoroquinolone class of drugs, which are vital in the fight against a whole range of potentially devastating bacteria including salmonella, E. coli and even MRSA. They are becoming less and less effective. It is estimated that one in eight cases of campylobacter poisoning is now resistant to ciproflaxacin. And the reason for that, it is believed, is because an antibiotic from the same group, enrofloxacin, has been used to treat chickens.

Enrofloxacin was first used in Holland in the 1980s but evidence began to emerge very soon afterwards that resistance was developing. In spite of that, its use was approved in this country in 1993. The inevitable happened. Resistance has increased in a variety of bugs to the extent that scientists at the Central Public Health Laboratory have warned of the risk it represents to human health. If he were still alive, the Japanese scientist to whom I referred earlier in this chapter would now be saying: 'I told you so.' As indeed he did.

There are now only two antibiotics used to make poultry grow faster: flavomycin and avilamycin. Their use is defended by the industry on the basis that they are not used to treat disease in humans and, therefore, will not weaken the resistance of antibiotics on which humans depend. Professor Pennington dismisses that argument. He says it really does not matter what drugs are fed to chickens or any other livestock because we simply do not know whether they will come in useful for treating human illnesses at some time in the future. The structure of avilamycin, for instance, is almost identical to an important new drug, ziracin, that went on trial in British hospitals to fight MRSA, VRE and strains of meningitis and pneumonia that are resistant to other antibiotics. Ziracin was withdrawn in May 2000. By a remarkable coincidence that was only seven days after the EU Scientific Steering Committee published a report which said avilamycin should be banned because of its possible threat to the effectiveness of ziracin. At about the same time the World Health Organisation accepted the principle that antibiotics should not be used for growth promotion if there was a medical equivalent. So if ziracin had been allowed to remain on the market there would almost certainly have been a worldwide ban on the hugely profitable avilamycin.

So what would Pennington do about drugs being fed to chickens to make them grow faster? He would phase them out altogether. He acknowledges that there will have to be radical changes made in farming practices, but the price of continuing is too high. He believes there is no alternative. It is difficult to find anyone in the medical profession who does not agree with him. Quite simply, too many bugs are developing resistance to too many drugs and the nightmare scenario is that we might run out of effective antibiotics. If that happens we will be back in the medical dark ages.

Many politicians share those concerns. But when it comes

to framing legislation and bringing in new regulations to control the use of antibiotics successive governments have, over the years, let us down badly. They have either failed to recognise the dangers or, once the evidence could no longer be ignored, they have given way to pressure from the industry. The noises from the poultry producers and the pharmaceutical manufacturers have drowned out those voices urging caution. When it has come to a choice between introducing strict new laws to control the abuse of antibiotics and caving in to the powerful industry lobbies, they have almost invariably caved in. By the turn of the century avilamycin was being fed to virtually every broiler chicken in the country.

The Netherthorpe Committee, back in the sixties, achieved virtually nothing. Hardly had the committee published its rather unhelpful report than it was clear that nature was fighting back in a worrying way. In farms across the country calves were falling sick with salmonella typhimurium which was not only resistant to the usual antibiotics but also carried extra-chromosomal genes, known as R-plasmids, which were capable of transferring to other species of bacteria within minutes of making contact. The government was persuaded that this was really serious and another committee – the Swann Committee – was set up. The committee reported the following year and recommended measures which seemed, at first sight, to be pretty tough. If they had been followed to the letter we would not be in the mess we are today. But they were not.

In theory farmers would no longer be allowed to use penicillin and the tetracyclines freely as growth promoters. They could use only drugs that would not 'impair the efficacy of a prescribed therapeutic antibiotic through the resistant strains of organisms'. In other words, drugs on which doctors relied to treat humans could no longer be fed to poultry or pigs as though they were Smarties. So far so good. But the legislation that followed left loopholes that were ruthlessly exploited. And successive governments, instead of forcing the industry to follow the spirit of the Swann recommendations, allowed it to get away with murder.

What the committee had done – whether it realised it or not – was to give a green light for the development of new antibiotics to be used purely as growth promoters.

In the years immediately following Swann the use of antibiotics on Britain's farms slowed down and even began to fall. But not for long. By 1977 it was rising again. Not only was there now a new range of growth-promoting antibiotics in use on Britain's farms, but poultry and animals were being fed more and more drugs just in case they fell ill. Professor Alan Linton, a member of the Veterinary Products Committee and now Emeritus Professor of Bacteriology at Bristol University, wrote a detailed paper in which he pointed out the dangers. He called it: 'Why Swann Has Failed.'

In the 1980s there were more serious outbreaks of salmonellosis in calves across the country. Again they carried the R-plasmids and this time they were resistant to no fewer than eight antibiotics used for treating disease in animals. Later those same bugs were transmitted to – and caused infections in – humans. As Professor Linton wrote in a preface to a report for the Soil Association in August 1999, the strains were capable of being genetically transformed into other types with even wider ranges of drug resistance. Swann had indeed failed.

In one sense it was hard to blame the farmers. Like any other businessmen, they wanted to increase their profits and produce as much meat as possible. They were getting an enormous amount of encouragement from Brussels. By then the EEC was handing out money like a drunken sailor. Some of the biggest wads of cash went to farmers who put up buildings specifically designed for intensive rearing.

You may wonder what the nation's vets were doing while all this was going on. They were facing a number of conflicting pressures. They had a professional duty to care for animals but they also had a duty to serve the interests of their clients, the farmers. If the food producers, supported by the law of the land, wanted to pursue a particular kind of husbandry, who were the vets to deny them? Many vets had their own concerns about the effect of growth promoters and prophylactic treatment on the resistance of bacteria but, again, they were paid to provide a service. Not that they even had to write out a prescription for growth promoters. The farmer could buy them when and where he chose. And there is an important difference between doctors and vets. Doctors are not allowed to sell the drugs that they prescribe. Vets are and they usually make a thirty per cent profit on sales. Those profits are vitally important to small rural vets, many of whom are struggling to stay in business. It is also important that vets should be able to carry a reasonable supply of drugs with them. If they're called out in the middle of the night to a farmer with a sick animal they can't give him a prescription and tell him to pop down the road to the nearest pharmacist. The temptation for an unscrupulous vet to sell drugs that he should not be supplying for a particular purpose is obvious. And, as in every profession, there is the occasional rotten apple.

In spite of the pressures, some vets did resist but often there was little point. If they told their clients they were not prepared to prescribe certain drugs for certain purposes the result, almost invariably, was that the farmers took their business elsewhere. If one vet refused to write a prescription, there was always another who had fewer scruples. There was enormous pressure from the pharmaceutical industry, too. Developing a new antibiotic is a horrendously expensive business. Alexander Fleming might have made his discovery with a dish and a bit of mould. To bring a new antibiotic to the market in the United States now can cost anything from $100 million to $350 million. You need to sell a lot of drugs to cover that sort of investment and it's not getting any cheaper as the problem of resistance grows.

The Swann recommendations, weakened though they had already been, were fought tooth and nail by the industry. The nation had to realise that these draconian measures could add three pence to the price of a pound of bacon! Outrageous! The very future of the British breakfast was at stake. If this bunch of meddlesome fellows got their way the poultry and pig producers would be driven out of business.

What Swann actually wanted was to ban certain important antibiotics from being sold without a prescription. In other words, they wanted to stop them from being used as growth promoters. He and his colleagues also wanted a single permanent committee to be set up which would oversee both the medical and veterinary use of antibiotics. It would be responsible, among other things, for monitoring the way in which antibiotics were used and how resistance to them was being affected.

In truth, that was the minimum the Swann Committee wanted. Originally they had hoped for more. They had watered down their recommendations because they had been told they would never get any tougher measures through. Even then they underestimated the power of the industry lobby and its determination to protect its growing market. Teams of lobbyists were despatched to Westminster to lobby the politicians. A sophisticated public relations campaign was mounted. Did the country not realise how much harm would be done to this vital industry if restrictions were to be imposed of the sort that Swann and his colleagues had recommended?

Instead of laughing that sort of self-serving nonsense out of court, the politicians caved in and allowed yet another antibiotic to be added to the list. It could be used by pig and poultry farmers for both therapeutic reasons and growth promotion. Remember how Swann wanted no growth promoters to have therapeutic uses? And then ministers went even further. Swann had been fiercely opposed to using growth-promoting antibiotics for adult and breeding cattle. Yet in 1976 the green light was given.

I had my first experience of drug-fattened cattle in the early eighties when I was invited by a farmer to admire the handsome beasts growing fat in his yard. Very nice, I said, but what about that thinner one in the corner? All the others had rear ends that looked like the biceps on an Olympic weightlifter, but this one looked like the 'before' bit in an old Charles Atlas advertisement.

'Something wrong with him?' I asked.

'Nothing at all,' said the farmer. 'That's the skinny bugger that's going to end up in our own freezer. The missus doesn't fancy the kids eating all those bloody drugs.'

So didn't he have any qualms about selling that beef to others?

'Look,' he told me. 'I've got a business to run and kids to bring up. It's hard enough making a profit out of beef as it is. The Ministry tells us what we can do and we do it. If we didn't we wouldn't be competitive and everyone would buy even more foreign beef because it would be so much cheaper. It's that simple. Anyway, they wouldn't let us do it if it wasn't safe. We just don't fancy it ourselves, that's all.' Quite so.

That farmer was injecting his cattle with hormones at the time and he was quite right: the government had assured us all they were safe. Then, a few years later they were banned in Europe. The British government initially opposed the ban, insisting the hormones were safe in the face of mounting evidence from European scientists, but in the end they had to submit to it. So the beef farmers switched to different drugs: those growth-promoting antibiotics.

It is the Veterinary Products Committee (VPC) which is responsible for deciding which drugs should be licensed. Until 1981 there was a sub-committee made up of microbiologists which advised the committee. They had been growing increasingly alarmed at what was happening with growth-promoting antibiotics but they had no powers of their own. They wanted to do as Swann had recommended: to review existing antibiotics and the way they were being used to see whether there should be more restrictions. One or two drugs were giving them particular cause for concern.

The chairman of the sub-committee, Sir James Howie, a former director of the Public Health Laboratory Service, wrote to the Minister of Agriculture and the Minister of Health. Their response was immediate. They refused to discuss the issue and, in 1981, the sub-committee was disbanded. By the end of the century the VPC still had no expert sub-committee of microbiologists dealing with this issue.

Nor do we have something else that Swann and every other concerned observer has called for over the decades: clear and precise details of all the antibiotics being sold to food producers across the country and the effect they are having. In Denmark there is a requirement to publish such information every year. In this country the distributors were asked if they would 'kindly compile sales figures'. There is no law to make them and no threat of punishment if they fail to do so.

Even the government's own advisory body which was set up in 1996, the Advisory Committee on the Microbial Safety of Food, acknowledges the problem. In a report in 1999 it said: 'It is very difficult to obtain the figures for the amount of antibiotics used in animal populations or of the number of animals treated.' It called for 'structured monitoring of the drugs used, the purpose for which they were used and the quantities concerned'. That has not happened. Under pressure from the committee and the Soil Association the government made a half-hearted attempt in May 2000 and published a few pages. At the time of writing there has still been no detailed report.

If successive governments showed extraordinary complacency while our defences against disease were being gradually eroded over the past fifty years there have been others prepared to sound the alarm. In 1997 the World Health Organisation drew attention to the growing crisis. In 1998 the House of Lords was sufficiently concerned to set up a sub-committee which held its own hearings. One expert witness after another traipsed into the faded splendour of the Palace of Westminster committee room to explain why, in their view, we were travelling a dangerous road.

In a powerful report at the end of the hearings the Lords spoke of a 'vicious circle repeatedly witnessed during the last half of the century, in which the value of each new antibiotic has been progressively eroded by resistance, leading to the introduction of a new and usually more expensive agent, only for this in its turn to suffer the same fate'.

In its chilling conclusion it warned of 'the dire prospects of revisiting the pre-antibiotic era'. It acknowledged that this will not happen overnight. It is a 'relatively slow but inexorable process, patchy in its effects but already under way. The options available for the treatment of infections have everywhere been constrained.' But 'in some locations the organisms causing several life-threatening infections are now resistant to all available antibiotics, so that for patients suffering these illnesses the antibiotics era has already ended'. Their inquiry, they said, had been 'an alarming experience, which leaves us convinced that resistance to antibiotics and other anti-infective agents constitutes a major threat to public health, and ought to be recognised as such more widely than it is at present'.

Perhaps the most graphic summary of the dangers we face was expressed by a witness, Dr Norman Simmons, who told the committee: 'It reminds me of the man who threw himself out of the Empire State Building and as he passed each window he said: "So far, so good." I know that we are out of the window. I just do not know how far we are above the ground.'

In some ways that old metaphor could be extended to apply to this whole book. Until the war we had a system of agriculture that was, by today's standards, inefficient. That's to say, it produced less food. On the other hand, it was kinder to the environment and to farm animals. We had no fears about pesticide residues or their effect on antibiotics or on the very earth beneath our feet or, indeed, on the fish in our seas and rivers. Years later we can see that we were taking a great gamble. Even those who defend intensive agriculture most fiercely acknowledge that it has produced problems. Where they differ from their opponents is that they say it was a price worth paying.

And now we are embarked on the greatest gamble of all: genetic modification. Where will that take us?

THE NEW GENE GENIE

GM

Welcome to the future and to the Brave New World of biotechnology. And what a wonderful world it has turned out to be. The setbacks suffered by the industry at the close of the twentieth century have long since been forgotten. The second generation of genetically modified crops overcame the concerns of the doubters and demolished the arguments of the opponents. They have delivered immense benefits for mankind, with the promise of even better to come.

This is a world in which plants provide medicines for many of our most terrible diseases. Hens lay eggs that contain cancer cures. Human vaccines that protect against hepatitis B are being grown in the humble potato. A human gene has been inserted into the tobacco plant so that the plant manufactures a hormone which will soon be used to cure Crohn's disease. Bananas are grown to produce a vaccine against diarrhoea. The pharmaceutical industry has a word for all this. It is called 'pharming'. The more conventional sort of farming is also reaping massive gains from advances in biotechnology.

Crops that once perished in poor weather now thrive. No longer must the world rely for the bulk of its food on good arable land with temperate climates. Food is grown in all but the world's most inhospitable regions – on the icy steppes and the wind-scorched deserts. The plants are able to survive the sharpest frost and the longest droughts. Even the waters of the oceans can now be used by farmers. Crops that would once have perished in saline soil have been genetically modified to flourish on land irrigated by salt water. The threat of water wars between drought-stricken countries has vanished.

Food that would once have rotted before it reached our dinner tables now stays fresh and wholesome. Not only does it taste better, it is better for our health. Many different varieties of 'functional' food have been developed, packed with all sorts of vitamins and minerals that nature never intended. Across the globe, many diseases that once thrived because of vitamin and mineral deficiencies are on the verge of being defeated. In the mid-1990s it was estimated that up to half a million small children went blind because of Vitamin A deficiency. Two-thirds of them died within a month of losing their sight. Now the risks are lower. Instead of the traditional rice eaten for generations they have what is called 'golden' rice, modified to contain both betacarotene and a healthy dose of iron. People with life-threatening allergies to nuts now eat them without fear. We can even enjoy a plate of chips without worrying too much about piling on the pounds. Potatoes have been modified to have a higher starch content which absorbs less fat.

The countryside in Britain is blooming. Powerful pesticides that were once applied repeatedly to control insects and weeds are used less and even, in some cases, not at all. Many of the vast boom sprayers that were such a familiar sight in the last century are quietly rusting away or are being cannibalised for spare parts. As a consequence, water sources that were polluted by chemicals are purer than they have been for many years. The plants themselves repel the insects that once devoured them. The weeds that choked their roots or starved young seedlings of nutrition and light can be killed with smaller doses of herbicides that cause no damage to the young crops themselves. Many plants have genes within them that alert the farmer to the first signs that they are under stress.

Harvests have never been more plentiful. A world in which billions faced starvation as the population increased relentlessly is now a world of plenty. No child need go to bed on an empty stomach. A wonderful world indeed. And everything I have described has been promised by those who promote GM technology. Some of it, they say, is happening now. Harvests are already improving and fewer chemicals are being used. If only we can rid ourselves of our fear of the unknown and embrace this new world, then anything is possible. As Churchill put it, the only thing we have to fear is fear itself.

Well, maybe. There is, of course, a darker scenario.

In this other Brave New World the introduction of genetically modified crops across the globe has led not to plenty but to even greater need. The promise of bumper harvests produced with fewer chemicals failed to materialise. The opposite happened. Pests have developed resistance to the poisons produced by the new GM crops and cross-pollination has produced weeds resistant to even the most powerful herbicides.

The whole balance of the global food economy has changed – with disastrous results for the world's poorest. Countless millions of peasant farmers can no longer afford to pay the bills of the agrochemical companies and are driven from their land to join the millions of other desperate people struggling to survive in the shanty towns or the gutters of the big cities. They even lost the right to grow food that they and their ancestors grew before them when patents were taken out around the world on different versions of 'terminator technology'. A terminator gene makes the plant sterile. The biotech companies said the technology was necessary to contain modified genes and stop them contaminating other plants. Their critics said it was to force farmers to buy new seed from them every year. In 1999 Monsanto made a 'public commitment' not to commercialise sterile seed techniques.

However, in a letter written six months later the chairman of the corporation, Robert B. Shapiro, said he did not rule out the future development of 'gene protection'.

The effect on small farmers was disastrous. Since the dawn of modern agriculture farmers have held back seeds from one year's harvest to sow the following year or to exchange for other seeds. They can do so no longer. In this new world where a tiny handful of immensely powerful multinational corporations are able to put their own price on the very stuff of life the poorest subsistence farmers suffer as they have never suffered before.

'Pharming' has failed. It has not filled the shelves of every doctor's dispensary with endless cheap drugs grown from the very same soil that produces our food, able to fight illnesses that were once incurable. Instead, new pathogens have been created by accident and threaten new diseases. There are no cures because no one properly understands what is happening. Even more threatening are the diseases created as a result of the food we eat. Most have been genetically modified one way or the other and many have produced new toxins. Some have created new allergies – all the more frightening because there is no way to test for them. Other genes have jumped from the micro-organism carrying them to others in our respiratory tract and gut and are seriously affecting the digestive system and breathing.

As for the new 'functional' foods, far from making us fitter and stronger they have caused enormous problems. It has proved impossible to keep a check on precisely what vitamins and minerals we are ingesting and many people are suffering from overdoses. The new food is also very expensive. In areas of the world where the poorest once relied on a little rice and a mixture of home-grown vegetables and herbs there is now severe malnourishment.

But a bigger crisis still is developing because of the

unpredicted consequence of creating new genetically modified crops. In the early years of the biotech revolution antibiotic-resistant genes were used in some genetically modified plants as a marker of genetic transformation. Fears were expressed at the time that the resistant genes might spread from the plant to micro-organisms with serious consequences for the treatment of disease. Those fears were ridiculed at the time but have since been borne out. Some genes that had crossed the barrier between one species and another without any apparent ill effects in the early stages have since resulted in frightening mutations. Strange things are happening to our digestive systems. New cancers are being reported. These are diseases for which, once again, there is no cure. There is no way of calling the transgenic material back. This brave new world is a place of nightmares – and no one can see how to banish them.

Those, then, are two possible scenarios – the best and the worst. There is an infinity of possibilities in between – neither as rosy as my first nor as frightening as my second. We may see some of the benefits of GM crops and some of the problems, but nothing we can't cope with. Or we may see public scepticism and growing concern eventually killing off the whole biotech dream. We have already witnessed in Britain the way the supermarkets reacted when public opinion was mobilised against what the tabloid newspapers dubbed 'Franken-foods'. Their customers made it clear that they did not want the stuff and the stores removed it from their shelves quicker than a store cleaner would mop up a broken egg. When the consumer speaks with such a loud voice, woe betide the retailer who does not listen.

Around the world there is opposition to the spread of GM, ranging from mild scepticism to outright bans. As I write no GM crops are being grown in Europe for commercial sale but we still eat the stuff. Almost three-quarters of all the GM food produced in the world is grown in America. The rest comes mostly from Argentina and Canada. More than half the soya beans grown in the States are genetically modified and a quarter of the corn is.

The driving force for GM technology comes from some of the most powerful corporations in the most powerful country in the world. And there is a disturbingly cosy relationship between those companies and the federal agencies who control and regulate the food industry. Some talk of a 'revolving door' between the big biotech corporations and American government departments. Every year more land is sown with GM crops, more patents registered by the biotech companies.

Throughout this book I have tried, whenever possible, to tap into my own experience. I have been a farmer for many years and have been able to learn a certain amount about conventional agriculture at first hand. I have tramped enough fields over the years to wear out several pairs of wellington boots. I have put on a diving suit to see the state of the sea bed beneath a cage on a salmon farm. Where I have no direct experience I have read a mountain of documents and books to seek the information I need. I have talked at length to experts on all sides of the debate – farmers, researchers, technicians, academics, doctors – and amidst a maelstrom of technical data I have even tried to apply a modicum of common sense.

Science, by and large, does not approve of common sense. Either something can be proved (or, rather, disproved) or it cannot. That's fine in a research laboratory but it's not much use in the kitchen. I cannot prove, for instance, that a certain amount of pesticide residue in my food will cause any specific harmful effect. Nor can anyone else; the necessary research has not been carried out and never will be. There are too many factors and influences involved. But that does not stop me using my own judgement when it comes to giving my small son a piece of fruit or a carrot. If it is a choice between food containing some residue of a synthetic chemical – however tiny – or food that is free of them, then I will choose the latter. I prefer not to take the risk, whatever the scientists may or may not say.

That approach is born out of a lifetime of listening to experts tell us there is no risk – only to discover years later that many of them were wrong all along. Sorry, they say, but now we know so much more about its effects, maybe it would have been better if that particular technology or that particular pesticide had never been approved for use. It may not be their fault. They may well have been acting in good faith at the time; they simply did not know enough. That is no consolation when it turns out that they got it wrong.

Many eminent scientists say we have nothing to fear from GM technology. They claim we have seen nothing go seriously wrong yet and the safeguards are so stringent we have nothing to fear in future. They denounce the 'green' organisations for using GM as a bogeyman to scare people into supporting their causes. Perhaps there is something in that; pressure groups by their nature make trouble. That's what they're for. But it's hardly as if the biotech companies don't have their own motives.

They have invested billions in this new technology and are investing billions more. They want to get that money back – and then some. If the world embraces GM food as they hope and plan, then the sky's the limit. The profits could be vast and the amount of control a small handful of companies would be able to exercise over the world's food supply could be near total. Farmers would have to buy their seed and their fertiliser and their pesticides from them. They would have no choice. Never, since prehistoric man sowed the first seed, have the stakes been so high. That is one reason why I describe this as the greatest food gamble of them all. And yet most of us know so little of what GM is all about or even precisely how it works.

GM propagandists have told us soothingly over the years that there is nothing terribly different between what they are doing and what plant and animal breeders have been doing for millennia. It's just a bit more sophisticated. I have lost count of the number of times I have been assured of that on the *Today* programme either by industry spokesmen or by government ministers who ought to have known better. It is simply not true.

With conventional breeding of the sort that has been practised for thousands of years different traits can be transferred only between plants or animals of the same species or one very closely related. Never before have we taken a gene from one species and inserted it into the DNA of another. In the words of the Food and Drug Administration in the United States, where GM technology enjoys enormous political support: 'The processes of genetic engineering and traditional breeding are different and, according to the technical experts in the agency, they lead to different risks.'

The genes come from all manner of species: perhaps from another plant, perhaps from a fish or even an insect such as a scorpion. They cannot just be sucked out of one species and inserted into another as if it were a blood transfusion. There has to be what is called a 'promoter'. That usually comes from potentially dangerous bacteria, viruses or other genetic parasites that can cause diseases. In the early days antibiotic-resistant genes were always used. What genes do is enable protein to be made – for good or ill. The proteins created as a result of the transplanted genes may be harmful because they have never been part of our food chain and are also new to the ecosystem.

To talk about 'inserting' a gene implies a degree of

precision. It conjures up pictures of a scientist in a white coat delicately injecting one microscopic organism at a precise point with genetic material taken from another. The reality could scarcely be more different. This is a random process. It is impossible to position the new gene with any accuracy in the DNA of the host plant. Scientists do not know where it will end up nor whether it will go in forwards or backwards nor how many copies will be included. Therefore the result is unpredictable; there is no way of knowing precisely how it will behave. So there is no way of knowing whether or not something may go horrendously wrong.

When that point was put to Professor Bevan Moseley, the former head of the Institute of Food Research and a member of the United Kingdom's Advisory Committee on Novel Foods and Processes, this was his answer: 'Well, I agree with you in the sense that when you use these methods you don't know what part of the chromosome the new gene is being introduced into and this is, you know, what I would say is a drawback to the technology.' Some drawback, you might think.

Professor Phillip James is the director of the world-famous Rowett Research Institute in Aberdeen and the man the government consulted when it decided to set up the Food Standards Agency. It was his report on which the agency was based. Here's what he said about GM technology: 'The perception that everything is totally straightforward and safe is utterly naïve. I don't think we fully understand the dimensions of what we're getting into here.'

But the GM supporters say: 'Show us what harm has been done by our technology and then we might listen to you. We've been marketing GM food of one sort or another for several years now and no one has suffered as a result.'

Can they prove that? No, they cannot. Nor can their opponents prove with absolute certainty that anyone has suffered. But some very serious doubts have been raised. If the last half-century of industrial agriculture has taught us anything, it is that we dismiss such doubts at our peril. If you added up all the GM food being grown at the end of 2000 it covered a total of almost 110 million acres – twice the size of the United Kingdom. It has been eaten by hundreds of millions of people. So you might assume that the regulatory process that led to its approval has been both thorough and as near failsafe as possible. That is what the biotech companies say. It is not true.

In 1990 three international organisations came together to discuss the potential risks presented by the biotech industry: the Food and Agricultural Organisation of the United Nations, the World Health Organisation and the Organisation for Economic Co-operation and Development. Representatives of the biotech industry attended the meeting. Consumer organisations did not. The industry faced a dilemma. They wanted international approval of their new foods and the easiest way to get it was to prove that they were not really so very different from traditional foods. But there were serious commercial risks involved in that. If there really was no difference between, say, GM soya beans and traditional soya beans how could they ever hope to prove that they owned the intellectual property rights in them and that their patents should be respected? The massive investment in GM technology was based on the assumption that there would be steady returns every time a farmer bought the company's seed. No patent: no profit.

But equally there was a danger in their creations being seen to be too different. If that happened they might have to clear steep hurdles when it came to getting approval for selling the food. What they most emphatically did not want was for GM foods to be treated as novel chemical compounds such as new medicines or food additives. If they were, the companies selling them would have to carry out a range of different toxicological tests for each new food. Not only would that cost a fortune, it would take a long time and delay their new foods coming to the marketplace – possibly by as much as five years. When a pharmaceutical company applies for a licence for a new drug it must submit enough data to fill sixty old-fashioned box files.

Even worse, the results of the tests could be used to restrict the amount of GM food we are allowed to eat. I referred earlier in this book to 'acceptable daily intakes' (ADIs) of certain compounds. If ADIs were set for novel GM foods they could never play more than a marginal part in our diet. It would be a commercial disaster and the industry was determined to avoid it at all costs. It succeeded. An agreement was reached which satisfied the biotech companies and the governments of the countries in which they operate and invest large amounts of money. The only people it did not satisfy were consumer organisations, pressure groups and the growing number of scientists who were uneasy about the uncritical acceptance of GM food. Depending on your point of view the agreement was either a masterly compromise or a disgraceful capitulation.

The agreement acknowledged that GM food is different, but only up to a point. So long as the new food was shown to be similar in its chemical composition to an equivalent food that had been developed using traditional methods, then it could be approved for sale to the public. Only if there were glaring differences might the food's manufacturer be forced to carry out more tests. That would be decided on a case-by-case basis. It became known as the concept of 'substantial equivalence'. In theory it sounds eminently sensible. In practice, many scientists believe it is worse than useless.

Take the case of soya beans that have been genetically

modified to make them resistant to the pesticide glyphosate. They are deemed to be substantially equivalent to traditional soya beans even though they are clearly different. If they were not, they would be killed when they are sprayed with glyphosate. The differences can be easily spotted in a laboratory. Ah yes, says the company that developed them, but there are more similarities than differences. But what happens when they are sprayed with glyphosate? Their chemical composition then changes significantly. And which beans were evaluated to determine their 'substantial equivalence'? Only beans that had not been sprayed. In other words, only those beans that would never find their way into the food we eat. It was not until some time after the absurdity of this was pointed out that the company belatedly agreed that perhaps sprayed beans should also be evaluated.

For more than a quarter of a century Dr Erik Millstone of Sussex University has been conducting research into the health and safety aspects of technological change in the food and agricultural sector. He is one of Britain's leading academic commentators in the field and no great admirer of the biotech industry. He believes that substantial equivalence is a pseudo-scientific concept because it is a commercial and political judgement masquerading as a scientific one. Indeed, he says, it is inherently anti-scientific because it provides an excuse for not requiring biochemical or toxicological tests. So it serves to discourage and inhibit potentially informative scientific research. Even on its own terms it has been wrongly applied within the regulatory process. He thinks the entire concept needs to be replaced with a practical approach which would actively investigate the safety and toxicity of GM foods rather than merely taking them for granted.

That is also the view of Dr Sue Mayer. Dr Mayer is a biologist who set up the policy research group Gene Watch UK. She shares the concern of many others that the interests of the biotech industry were put above public health when the regulations were drawn up. She also believes that the intention of substantial equivalence is political – to reassure consumers, avoid labelling and maximise profit – rather than making a serious assessment of the possible hazards. She wants to see a new system. First, she says, we should ask whether the GM food is needed. Then we should look at the alternatives and decide whether they are less risky. And only then, if these hurdles are passed, should we go on to a comprehensive assessment which would take difference and not similarity as its starting point.

The British government has tacitly acknowledged the concerns. The Food Safety Minister, Jeff Rooker, spoke approvingly in the House of Commons in July 1998 about new, genetically engineered 'wonderdrugs' and he went on: 'There is some comfort in the regulatory process for medicine which, I admit, is not in place for food and agriculture.' The following year an internal memo written by the then Cabinet Office Minister, Jack Cunningham, about GM crops and foods was leaked. In it Dr Cunningham asked: 'Why don't we require a pharmaceutical-type analysis of the safety of these foods with proper trials?'

So what sort of things should be looked out for? According to the government's own chief scientific officer and chief medical officer there are half a dozen ways in which the genetic modification of food could, in theory, affect human health:

- The inserted gene itself may have adverse affects.
- The inserted gene may lead to the production of a protein that is toxic to human beings or that may produce an allergic reaction. Genes are the blueprints that the body uses to manufacture protein. Protein is what makes up every cell in our body.

- The inserted gene may alter the way that existing genes in a plant or an animal express themselves. That could, in turn, increase the production of existing toxins or it could 'switch on' genes that had, until now, been 'silent'.

- The inserted gene may alter the behaviour of a micro-organism that has been used to carry it into the host plant. The micro-organism might have been harmless in its original state but it could change as a result of the new gene it is carrying.

- The micro-organism that carries the new gene might alter the balance of all the other micro-organisms in the human gut.

- The inserted gene might go walkabout. Instead of staying locked into the micro-organism that is meant to be carrying it, it could be transferred to others in the human gut or the respiratory tract. Even if it were perfectly innocuous when it was married to its intended partner it could be harmful if it joined up with others.

Professor Terje Traavik is particularly worried about that last one. He is the head of the department of virology at the University of Tromsø in Norway and for nearly thirty years he has been conducting research in the fields of viral and gene ecology. He spoke about the risk of 'horizontal gene transfer' when he gave evidence in July 1999 at the trial of twenty-eight Greenpeace activists who had destroyed a crop of GM maize at a farm in Norfolk. It means the genetically engineered DNA being taken up directly into the cells of unrelated species and incorporated into a cell's genome. It might then be transferred to bacteria and viruses with unknown consequences.

Traavik said that once a horizontal gene 'cascade' has begun there is no way of stopping it. It may take a long time for researchers even to become aware that it has started. Like so many scientists working in this field Traavik admits the difficulty of making any realistic risk assessment simply because we do not know enough about the way genes behave when they are taken from one species and implanted in another. Nor are there any reliable methods available to detect and monitor any aberrations. But what he did say was deeply worrying: 'Horizontal gene transfer from GMOs [Genetically Modified Organisms] is a real option. Such events may result in extensive and unpredictable health, environmental and socio-economic problems. Under some circumstances the consequences may be catastrophic.'

Scientists like Mayer, Millstone and Traavik are not alone in their worries. The Institute of Science in Society – a radical, campaigning organisation – invited scientists from around the world to sign a petition calling for an immediate moratorium on further environmental releases of transgenic crops, food and animal-feed products for at least five years. The petition also called for a ban on patents on living organisms, cell lines and genes and a 'comprehensive, independent public inquiry into the future of agriculture and food security for all, taking account of the full range of scientific findings as well as socio-economic and ethical implications'. By December 2000, the Institute said, its petition had been signed by 378 scientists from forty-eight countries.

The director of the Institute is Dr Mae-Wan Ho, a visiting reader in biology at the Open University, who began research on human biochemical genetics at the University of California at San Diego in 1967. She believes that GM is an untried, inadequately researched technology and that there is no evidence capable of standing up to scientific scrutiny that GM crops are safe. On the contrary, she says, the entire technology is inherently dangerous because it involves making artificial genes and combinations of genes that are transferred into cells and embryos to create GMOs, none of which may ever have existed in billions of years of evolution.

Equally, there are many scientists who say the risks are grossly overstated or do not exist at all. Professor Don Grierson is the head of the plant science section at Nottingham University and the man credited with creating the genetically modified tomato with a longer shelf life. When I asked Monsanto to recommend a scientist to speak on behalf of GM technology it was his name they offered. I asked Professor Grierson why so many of his academic colleagues had so many misgivings. His answer was succinct: 'Because they ask the wrong questions.'

Professor John Beringer is the Dean of Science and Professor of Molecular Genetics at Bristol University and was both the chairman of the Advisory Committee on Releases to the Environment and a member of the Advisory Committee on Genetic Modification. In an article for the journal *Biologist* he asked: 'What is the future for GMOs?' He answered his own question. 'Excellent.' Professor Beringer believes it is inconceivable that 'even in twenty years' time people will challenge the idea that GM technology is vital for human health and happiness'.

It is not unusual for scientists to disagree. Indeed, since science is nothing more than the search for knowledge it would be remarkable if they did not. But what is happening here seems to me to be quite extraordinary. On one side of the debate we have experienced and well-respected scientists warning of the most terrible consequences for our health if we pursue this technology as we are doing at the moment. On the other we have equally experienced and equally respected scientists such as Professor Beringer telling us it is vital to our future health and happiness that we should do precisely that. Note that word 'vital'. He does not say 'helpful' or 'desirable' or even 'important'. He says it is vital. Opponents of GM are often called hysterical and accused of being anti-science when they use strong language to argue their case. You might wonder how Professor Beringer explains the occasional outbreaks of happiness the human race has managed to experience since we first crawled out of the primeval slime.

But this is no arcane dispute conducted on the internet or the pages of the scientific journals that we laymen can afford to ignore until the scientists have reached a consensus and can tell us what to think. Every day in laboratories around the world researchers are looking for new ways to manipulate genes and transfer them from one species to another. The Monsanto laboratories in St Louis employ a vast number of researchers peering into microscopes working on modifying cells to develop seeds with specific characteristics. Every day more GM crops are grown. According to the biotech companies the acreage planted in 2000 was almost eleven per cent more than the year before. So instead of looking into the future let us look back at what has already happened. Does the history of GM technology inspire confidence or fear?

Left to her own devices an animal will produce enough milk for her offspring with, ideally, a little to spare for emergencies. When the youngster can feed himself her milk will dry up until she produces another little hungry mouth. Farming changed all that. After years of selective breeding and carefully planned diets, a healthy dairy cow was able to produce far more milk than she needed for her calf. By the 1950s some were delivering 4,000 litres a year. But it still wasn't enough for farmers tempted by high prices and guaranteed sales. So they tried different feeds, ever richer in proteins, and silage instead of hay, and even more selective breeding. In twenty years the yields of many herds doubled. Some cows were delivering as much as 10,000 litres a year. By the 1990s cows were so highly bred to gush milk that if they did not eat abnormally large amounts of protein their body condition would deteriorate dramatically. It became impossible, in many cases, to feed the most high-yielding cows enough food to maintain them in good health.

Whether they knew it or not, farmers had been feeding their animals bits of other cows in their concentrated food pellets. Some of that food was, it is now believed, responsible for the tragedy of BSE and variant CJD. It was a high price to pay for a bigger milk yield.

In the laboratories of the American biotech companies they were working on a different approach to increase yields. The Russians had been the first to come up with the idea back in the 1930s. They discovered that it was possible to isolate a growth hormone known as bovine somatotrophin (BST) or bovine growth hormone (BGH) in the pituitary gland of a dead cow. If that hormone were injected into a live cow she would produce much more milk. The problem was that it cost so much time and effort to obtain the hormone that it made no economic sense, so they shrugged their shoulders and got on with other things. Half a century later, Crick and Watson made their Nobel Prize-winning discovery that unlocked the secret of DNA. From then on genetic engineering was possible. It meant that an endless supply of the hormone could be produced at very little cost.

By 1990 the Americans had cracked it: a genetically engineered hormone called recombinant bovine growth hormone or rBGH. Right from the beginning the US Food and Drug Administration, a powerful supporter of genetic modification, was behind it. By 1993 they had given the Monsanto Corporation permission for it to be marketed and within a very short time a third of America's vast dairy herd was being given weekly injections. Virtually every American citizen was exposed to rBGH in some form or another: milk, cheese, yoghurt or ice cream.

There were, apparently, no ill effects. True, the cows

themselves suffered. They developed infections far more frequently and had to be treated more often with antibiotics. But there seemed to be no harm to the humans who drank their milk. How curious, then, that Monsanto could persuade no other country it approached to approve the sale of this new hormone for use on its dairy farms. After all, here was a relatively cheap and easy way of producing more milk which had been sanctioned not only by the FDA but also by the World Health Organisation body, the Codex Alimentarius. In fact, there was nothing curious about it all.

Canada was one of the countries that said no. A government report released a few years later showed why. Canadian government scientists took a close look at the findings of a study conducted by Monsanto, which had been reported by the FDA in the journal *Science*, and flatly contradicted it. The FDA had said that rats fed high doses of rBGH over a period of ninety days had suffered no ill effects. The Canadian scientists said they had. Nearly a third of the rats developed responses which showed that the rBGH had been absorbed into their blood. Cysts were reported to have developed on the thyroids of the male rats and the stuff got into the prostate gland too. One FDA official told the Associated Press in 1998 that they had, in fact, never examined the raw data from the study but had based their conclusion only on a summary.

The Canadian scientists produced their own report in which they said: 'The usually required long-term toxicology studies to ascertain human safety were not conducted. Hence, such possibilities and potential as sterility, infertility, birth defects, cancer and immunological derangements were not addressed.' Years later an inquiry was held into the whole episode. The scientists who wrote that report testified that they had come under pressure from government officials to alter its content before it was published. Two of the authors and four other Canadian government scientists testified that they had been threatened by government officials with transfers to other jobs where 'they would never be heard of again' if they did not speed up the approval of rBGH in Canada. A Canadian legislator, Mira Spivak, served on a committee investigating the approval process. She said her staff had been given a copy of the study in which the information about the potentially troubling effects of rBGH on rats had been 'blocked out'. To this day rBGH has not been approved for use in Canada.

It met the same fate in Europe. The European Commission instructed its scientific committee to look into what might happen to humans who drank milk from cows injected with the genetically engineered hormone. The committee has sixteen scientists on it from different countries in the European Union and is regarded as both independent and authoritative. It found that when cows were injected with rBGH they produced excess levels of something called IGF-1, which is Insulin-like Growth Factor-1. It occurs naturally in cows but excess levels pose serious risks of breast and prostate cancer in humans. The report concluded: 'Experimental evidence for an association between IGF-1 and breast and prostate cancer is supported by epidemiological evidence arising from recently published cohort studies.' It warned that excess levels could promote the 'growth and invasiveness of any cancer by inhibiting the programmed self-destruction of cancer cells'. There was also a problem with antibiotics. Cows treated with rBGH to produce more milk suffer more attacks of mastitis. Hence more antibiotics are used and the danger from residues grows.

So what are we to conclude from all this? Well, what cannot be proved is that anyone has been harmed by eating yoghurt or drinking milk from a cow treated with rBGH. It's true that the cancer statistics in the United States show a steady rise in the level of breast and prostate cancers. There are a million new cases of cancer reported in America every year of all kinds. It is easy to make an assumption and assume a link. But that's all it would be: an assumption.

There might be any one of a dozen reasons why the incidence of cancer, and other diseases, has been increasing. Some of us are more prone to certain illnesses than others. It may be because of genetic predisposition. It may be because of the way we were brought up. It may be because of a thousand different reasons that we simply do not yet understand. The skinny little mite reared in a slum on a diet of burgers and chips, who spends his babyhood crawling around on a filthy floor shared with mangy cats and scabrous dogs, may grow up to be a healthy hunk who enjoys rude good health until he keels over with a heart attack at the age of ninety. The middle-class child fed a healthy, balanced diet and ferociously protected from germs may become a wheezy asthmatic forced to take to his bed the moment someone sneezes within ten yards of him. And we don't know why. In the GM future genetically engineered hormones might play a big part or a small part or none at all. The only thing we can do is watch and wait.

But even if degenerative diseases increase at a still faster rate over the coming years it will be virtually impossible to prove a causal link. When an entire population is subjected to something, how do you measure its effect? Where is the control group with which the 'affected' section of the population can be compared? There is none. And even if it could be shown that there are more cases of breast and colon cancer in the United States than there are in any other country it would still prove nothing. There could be other lifestyle or environmental reasons peculiar to America to explain the increase. Unless there were to be a massive epidemic or some terrible new disease that coincided with the introduction of rBGH milk, which could realistically be explained by nothing else, then no blame could be attached with any certainty to the genetically modified hormone. The same can be said of all the other GM food now being grown on farms around the world.

A vast number of people are eating genetically modified food – mostly maize and soya beans – and it seems that there have been no harmful effects of any kind. The conclusion, say the biotech companies, is obvious: GM food is safe food. Can they prove that? No.

The reason they cannot is rooted partly in the same logic that makes it almost impossible to prove that GM food is not safe. If almost the entire population is eating food that contains GM ingredients and there is an increase in the number of people suffering from, say, allergies of one kind or another or gastrointestinal disorders, how do you prove that GM is to blame? How, for that matter, do you prove there has been an increase in the first place?

In this age of massive databases, in which any credit card company can tell you how much you spent on that cheap weekend in Paris and whether you could afford it, the amount of information stored on our state of health is surprisingly limited. We monitor deaths, births, abnormal labour, cancer cases, food poisoning and a few others but that leaves a vast range of illnesses for which there are simply no records. Gastrointestinal illnesses are just one of them. When it comes to monitoring what we eat as a nation and the effect it might or might not have on us, we know almost nothing.

But even if the statistics were available and we knew there had been an increase in this or that disease it would be virtually impossible to prove a link with any particular food. We are subject to a vast range of environmental influences. Every time we come down with a touch of gastric trouble someone, somewhere, always says sympathetically, 'Ah yes, there's a lot of it about.' But we never really know what 'it' is and, unless we're on our deathbed, the doctor usually shrugs his shoulders and tells us to stop eating, drink plenty of water and get lots of rest. The truth is, he doesn't know either. So we might have our suspicions. But suspicion is one thing, proof is quite another.

We each eat many tons of food in a lifetime and we and the food manufacturers prepare it in a thousand different ways. Some of that food might have a large GM content; some will have none. The British Medical Association wants all GM products to be segregated at source so that they can be identified and traced. It says that if the biotech industry persists in mixing GM and non-GM foods in the same product, the Food Standards Agency should consider banning such products from this country until any possible hazards have been eliminated. Needless to say, the United States objects strongly to a ban. It goes further in wanting no labelling of any kind, neither to identify GM foods nor to identify foods that have no GM content at all.

Because of the concept of substantial equivalence there is minimal testing before a new GM food is approved. There is no strict, legal requirement for long-term testing, for testing on humans or for specific dangers to children, or for testing for effects on people who are allergic in different ways. Certainly there are problems with some testing. Scientists need to know what they are looking for in order to test for it. If a new genetic combination produces some unexpected toxins or allergens there is no way of testing for them in advance precisely because they are unexpected. So is there anything we can point to with absolute certainty and say that it has gone seriously wrong and people have suffered as a result? Specifically, has anyone ever died because of GM? The answer to that, say many of its opponents, is yes. And the food involved is a supplement most people have never heard of. It is called tryptophan.

Food supplements are often made in vats, using a fermentation process. Bacteria are grown in the vats and the supplements, such as amino acids, are extracted and purified. One of those amino acids – tryptophan – has been produced in that way for many years. Then, in the 1980s, a Japanese company called Showa Denko decided to try something different. Their scientists used genetic engineering to change the bacteria in a way that would make the process more efficient. It worked. There was a substantial increase in the production of tryptophan and it went on sale in the United States in 1988. The company was not required to conduct any safety tests because tryptophan had already been shown to be safe and no one was interested in the changes in the manufacturing process.

Within a few months of the 'new' tryptophan being sold thirty-seven people who took it had died and about 1,500 people were permanently disabled. It took months more to discover that the reason was a toxin in the genetically engineered tryptophan that caused a disease known as EMS. Those are the bald facts of the case and they are not disputed. What has remained in dispute ever since that terrible tragedy is whether the toxins were produced as a direct result of genetically modifying the bacteria used in the manufacturing process.

One explanation, favoured by the biotech companies, is that the company had cut corners with its filtration procedures at precisely the time it switched to the GM bacteria and thus allowed impurities to pass through. The lawyer representing the company said its own scientists dismissed that explanation because they had changed their filtration procedures before in much the same way without any ill effects. There are two other significant points. The lethal toxin that caused so much death and suffering had never been found by Showa Denko in bacteria that had not been genetically engineered. Nor had any other company producing tryptophan ever caused an outbreak of EMS.

Opponents of genetic engineering are in little doubt that the GM bacteria was the cause of the deaths. We shall probably never know for sure. Showa Denko has never released the genetically engineered bacteria strain. They say they destroyed all stocks of them when the tragedy began to emerge.

Fear of allergies became the new disease of the late twentieth century. We worry about them more now than we have ever done and often with good reason. Take nuts. Most of us eat them dozens of times a year in one form or another. They are used in all sorts of food and if we haven't bothered to read the label we won't even know we're eating them. That's no problem – unless we are allergic. In that case one bite can be fatal. The lives of many parents are dominated by a single imperative: to protect their highly allergic children from nuts at all costs. I know one couple who moved house and paid fees they could ill afford to send their little girl to a private school where there was always a nurse on duty just in case she ate something she should not and suffered an allergic reaction. Her allergy is so severe that she'd probably be dead by the time an ambulance could get her to hospital.

Most allergic reactions are, mercifully, less dramatic but they can still be serious, particularly in children and old people. There is evidence that genetically modified foods may result in new allergies, all the more worrying because it takes years for new allergies to become apparent and for detectable increases in existing allergies to be picked up. But what has already been proved beyond doubt is that transferring genes from one plant species to another can make the host plant allergenic in a way that it was not before. The genetic engineering in this case involved transferring a gene from brazil nuts to soya beans.

One of the problems with soya beans is that they are not

nutritious enough. Their protein is deficient in an amino acid called methionine. That is a serious problem for animal feed and for humans who eat a wholly vegetarian diet. Over the years various attempts have been made using traditional breeding methods to manipulate the balance of the soya bean's amino acids, but improvements in nutritional quality either failed or came at the expense of yield or the quality of the crop. Brazil nuts have plenty of methionine. So the idea was to transfer a gene from the nut to the bean. You would then have a soya bean with a much higher nutritional value and no need to use a supplement.

Just as with tryptophan, it seemed to work. But then a group of five scientists and doctors carried out tests on the GM soya bean to see what else had happened. They reported their results in the *New England Journal of Medicine* on 14 March 1996. Apart from being highly nutritious, brazil nuts are also dangerous to anyone with a nut allergy. What the researchers had found was that the allergenic property of the nuts had been transferred to the beans. The dream of a more nutritious soya bean had been destroyed.

The GM industry cites this as an example to prove how thorough is the testing process. Proper tests were carried out and as a result the transgenic food was found to be seriously flawed. No damage was done, except to the bottom line of the company involved, and that was the end of it. But that tells only part of the story. Here was a team of qualified people, supported by a grant to meet their costs, who knew precisely what they were looking for. The brazil nut allergen was well known and could be specifically tested for. They had samples of blood from people with an established allergy to help them with their tests. In other cases, of course, the allergen might not be known. It is entirely possible that its effects might appear only over a period of time. It might produce a form of allergy of which we have no experience. It might simply not be identified. Far from exonerating the industry, what this little tale tells is that the risks exist.

The millions of acres of GM crops grown around the world have been modified not to make them more nutritious but for more prosaic reasons. The vast majority have had new genes inserted to make them resistant to powerful pesticides or to stop pests eating them. Some scientists claim that there are unexpectedly high levels of toxins and allergens in them. The industry says the variations are within the normal range you would expect to find. The problem is that because of the problems with testing it is impossible to establish what variations are acceptable. There have been reports of increases in allergies in the United States, but nothing has been proved.

The other main GM crop is maize. Some of it has been grown for human consumption, most of it for animals. The biotech company Aventis developed a maize that produces an insecticide protein known as Cry9C. They called the maize StarLink. It was approved for sale only as animal feed but something went wrong. In 1999 Aventis could not account for all the StarLink corn grown by American farmers. Suspicions were aroused and when checks were carried out it was discovered that the corn had been used in food that went on sale in American supermarkets for human consumption. All the food was recalled and Aventis was forced to offer to buy back every bushel of maize grown on a third of a million acres of land. There are two lawsuits against them and some estimates put the total cost of the fiasco at a billion dollars. Now the suspicion is that the regulatory authorities in the United States will find a way of giving StarLink retrospective approval as food fit for humans.

The fear of many scientists and organisations such as Greenpeace is that Cry9C can cause dangerous allergic reactions in some people. The American Environmental Protection Agency used more cautious language but when it reviewed the research data on StarLink it found it to be either inconclusive or indicating that 'Cry9C exhibits some characteristics of known allergens'. Its advisory panel was told by some scientists that there is no known safe level of allergens in food. After that report was issued Aventis did some more research which, it said, showed there was no risk to consumers from the potential presence of very low levels of StarLink corn in food. Sceptics wondered why, if that were indeed the case, the research had not been carried out before the crisis broke.

On this side of the Atlantic the star of the most celebrated GM scare story was Dr Arpad Pusztai of the Rowett Research Institute. In April 1998 he went on television to announce that he had conducted experiments on genetically modified potatoes. They showed, he said, that rats who had eaten the potatoes had suffered serious intestinal changes. There had been alterations in the structure and function of their digestive systems, changes in the normal development of their internal vital organs and a slowing down of their immune responses to outside injuries. Most importantly, he said, those changes were unpredictable. He himself would not dream of eating the potatoes and it was 'very, very unfair to use our fellow citizens as guinea pigs'. This, it seemed to many, was the smoking gun, the evidence that GM food was dangerous to our health.

The British Medical Association called for a moratorium on planting genetically modified crops. Within two months seven chains of supermarkets across Europe had announced they would clear their shelves of GM food and sell no more. Three of the world's biggest multinational food companies said they would not sell GM food either. The Rowett Institute responded by sacking Dr Pusztai for breaking the rules of the Institute and publishing his research before it had been completed or peer-reviewed. He was publicly defended by many scientists and savaged by others.

The Royal Society reviewed what it could of the research conducted by Pusztai and his colleagues and pronounced that they had found it to be flawed. The *Lancet*, outraged by the behaviour of the Royal Society, decided to print Pusztai's work. In a fiery editorial it said the Society had been guilty of a 'gesture of breathtaking impertinence'. The researchers should have been judged 'only on the full and final publication of their work'.

Pusztai, now in retirement, has remained defiant in the face of the attacks on his reputation. He believes that the consequences of genetic manipulation of our food might hold untold dangers for humanity. Like many others, he is appalled that inadequate testing methods have been accepted at face value by the regulatory authorities. He believes they are not remotely sensitive enough to detect changes that might occur in our bodies as a result of eating GM foods.

You will have noticed how many times the words 'might' or 'may' crop up in any assessment of the risks involved in genetic manipulation. You will also have noticed how divided are the scientists. How can we ordinary mortals be expected to make our own assessment when the men and women with white coats and a language we cannot speak argue so fiercely among themselves? We cannot. But what we can do is to define risk.

The mistake that is often made is to confuse 'risk' with 'probability'. It may well be that the probability of something unpleasant happening is great but the consequences of it are insignificant. If you walk about barefoot the probability of stubbing your toe against the leg of your bed at some time in the next year may be great. But you are hardly likely to spend too much time worrying about it and you'd be considered pretty crazy if you got rid of the bed and slept on the floor to make certain it did not happen.

If you smoke twenty cigarettes a day the probability that you will get cancer or have a heart attack in the next year will be smaller but the consequences will be much greater. It is even possible that it will kill you. So you would be considered eminently sensible if you decided to throw away your cigarettes immediately and never smoked again.

In both cases – consciously or unconsciously – you would have arrived at the risk by multiplying the probability of a particular event occurring by the consequences of that event. Getting a stubbed toe is no great risk; getting cancer is an enormous one. Therefore the risk of smoking is unacceptable. Unless, that is, you derive so much pleasure from doing so that you really don't want to live if you can't smoke. That's the other factor in the risk equation: benefit. Just about everyone who uses a mobile phone knows there is the possibility that it may be doing them some harm. One of the biggest phone companies in the world is being sued for billions by people claiming their brains have been damaged. Yet most of us keep using the wretched things. And the reason is simple: we reckon that they bring us great benefit.

So those, it seems to me, are the three factors that we ought to weigh up when we consider the risks posed by genetic food: probability, consequences and benefit.

The first area – probability – is the trickiest. We simply do not know how likely it is that something will go wrong. We cannot even be certain whether or not something has already gone wrong. It might be that the GM food we have eaten so far has caused subtle changes to our bodies that has already harmed our health and the more we eat the worse it will get. It might be that it has had no effect at all. Many scientists believe the probability of something going badly wrong is enormous. Others say it is negligible. Others say it is non-existent.

The second area – consequences – is easier to deal with. At

the beginning of this chapter I created an imaginary nightmare scenario. It was easy to find the material. A quick trawl through academic papers and comments posted on the internet produces a seemingly endless list of scientists, some of them very distinguished indeed, who fear the worst. Here's a tiny sample:

Dr George Wald, Professor Emeritus in Biology at Harvard and a Nobel Prize-winner in medicine, thinks GM technology could 'breed new animal and plant diseases, new sources of cancer, novel epidemics'.

Dr Samuel Epstein, Professor of Environmental Medicine at the University of Illinois, is 'sure that there will be a significant increase in deaths from certain types of cancer. If that is the only adverse effect we will have been lucky.'

Dr Vyvyan Howard, an expert in infant toxico-pathology at Liverpool University, believes that 'swapping genes between organisms can produce unknown toxic effects and allergies that are most likely to affect children'.

Dr Mae-Wan Ho believes they could 'unleash pandemics' of infectious diseases that are resistant to antibiotics.

The one great fear most GM sceptics have in common is that if something does go badly wrong it may be impossible to put it right again. This is a genie that we may not be able to put back into the bottle. So in terms of the risk equation, the probability is uncertain and the possible consequences are disastrous. What about the benefits?

The glossy leaflets printed by the biotech companies are packed with happy, smiling peasants, backlit photographs of pretty mothers bouncing chubby babies on their knees and many promises. The reader is told that biotechnology can improve the quality, taste and nutritional benefits of food. It has not happened yet and the evidence so far suggests that those claims are overstated. But the big promise is – and always has been – that biotechnology can feed the world.

Without it more people will go hungry as the population increases. If that is true, then clearly it alters the whole risk equation. The benefit would be of incalculable value. Who could condemn millions to a future of starvation if GM can feed them? But it is not true.

For all the talk of the world's population exploding in the next century, the reality is different. The latest assessment of population trends by the United Nations indicates that there will be a 'drastic deceleration' in growth. It peaked in the second half of the 1960s at 2.1 per cent a year. By 2050 it will have fallen to 0.3 per cent. Over the next thirty years crop production in the developing countries is projected to be seventy per cent higher than it was in the late 1990s. That projection ignores the impact of GM technology. The FAO report says that for the world as a whole 'there is enough, or more than enough, food production potential to meet the growth of effective demand [. . .] that is, the demand for food of those who can afford to pay farmers to produce it'.

The key phrase here is 'can afford to pay'. There are three reasons for hunger in the world today: war, problems with distribution and, overwhelmingly, poverty. There have been terrible famines in the two biggest countries in the world – China and India – at times when there was a surplus of food. It is estimated that there are 1.3 billion people living in real poverty in the developing world. Of those 840 million suffer from malnutrition. People who can afford to pay do not go hungry. The biotech companies say that GM technology will feed the world's poorest. Instead, it could have the opposite effect.

In 1999 the charity Christian Aid produced a report that said GM crops were creating 'classic preconditions for hunger and famine'. Over the past few years the multinational biotech companies have spent a fortune buying up many of the biggest seed companies around the world and patenting the different varieties of seed. Christian Aid said that a food supply based on too few varieties of patented crops is the worst option for food security. Some say that poor countries will embrace the new technology because it will enable them to grow enough to feed themselves. History tells us that they will end up growing cash crops for sale to foreign countries to earn the money to pay for the technology.

So that is the risk equation. The probability of something going wrong is uncertain. The consequences of something going wrong are devastating. The benefits have, on the evidence so far, been grossly exaggerated. Dr James D. Watson, the scientist who co-discovered the double helix structure of DNA, won the Nobel Prize for it and made genetic engineering possible, says it is a matter far too important to be left solely in the hands of the scientific and medical communities. Another Nobel Prize-winner, Professor George Wald of Harvard University, believes this is the largest ethical problem science has ever had to face and he warns that everything is happening 'too big and too fast'. He says that 'restructuring nature was not part of the bargain . . . going ahead in this direction may be not only unwise but dangerous'.

THE COUNTER REVOLUTION

Organics

When the Chinese leader, Chou En-lai, was asked for his assessment of the effect of the French Revolution he paused for a moment and then said: 'It's too early to tell.' He may have been a shade too cautious, but you can see his point; history takes a long time to settle down. We think a particular action will have a particular consequence and years later, for a hundred reasons we had not even considered at the time, we discover we were wrong. There was a revolution in agriculture after the war – the birth of factory farming and industrial agriculture – and we are still assessing its effects because they continue to this day. But there has also been a counter revolution of sorts in the past fifteen years in the food industry. It is going to take a very long time before we can be sure of its effects because it, too, is still in progress. It is the organic revolution.

In 1981 it was just about possible to buy organic food in a Safeway supermarket – assuming you knew there was such a thing and assuming you wanted it very badly indeed. Four years later if you looked hard enough in one or two Sainsbury's or Waitrose stores you might find a few bruised apples or limp carrots with bits of soil still clinging to them on sale at silly prices. Set alongside the regimented rows of 'conventional' food – all bright and shiny and uniform and packed in cellophane and looking as fresh as the blush on a new bride's face – they were more than a little sad. You wondered why the supermarket bothered and you wondered who would dream of buying the stuff. The fact that it was there at all seemed no more than a gesture – perhaps a toe in the water to see if there could conceivably be a market which they might one day exploit to make a pound or two of extra profit.

Most supermarket managers looked on with a mixture of amusement and contempt. They said the days of muck and mystery had disappeared a long time ago and shoppers wanted their fruit and vegetables to be pristine, uniform and cheap. They did not want their apples to have blemishes on their skin; they had to be clean and shiny. They did not want some carrots to be long and thin and others to be short and fat; they must all look much the same. Some supermarket bosses said sniffily that the quality of organic food could not be guaranteed. And some said: why bother? We're doing very well as we are. But ten years later all the supermarkets were taking a serious interest. Within another five years they were fighting each other for a share of a market that was growing faster than any other sector in the industry. Instead of dipping their toes in the water they were jumping in fully clothed, desperately trying to grab their share and drown their competitors. So what had happened to change things so dramatically?

It's easy enough to say what did not happen.

There was no encouragement from our political leaders. They either did not know what was beginning to happen or they did not care. There was no great marketing drive by the big retailers. That's what normally leads to any real change in our buying habits. Any marketing expert will tell you that it's not difficult to persuade us to buy things if the budget is big enough. All the big retailers and food producers spend a fortune on advertising and marketing to keep us buying certain brands or to get us to start buying new ones. Ask Coca-Cola or McDonald's. But there were no full-page advertisements in the national press, no flashy thirty-second commercials with glamorous models in the advertising breaks on ITV. Not until many years later did the retailers start using organic food in their advertisements to tempt fickle shoppers. In terms of marketing, this was the quiet revolution.

As even Chou En-lai would concede, revolutions can succeed only if they have the support of a significant section of the population. They do not receive that support unless the people are unhappy with the status quo. So it was with food. There were enough people becoming a little uneasy about the quality and safety of the food they were buying for the supermarkets to spot that something was starting to happen. The reasons for the unease were mixed. Some people were worried about the use of pesticides. Some were offended by the worst excesses of factory farming and the effects it was having on the animals who suffered under it. Some were nervous about the growing number of food scares. And when those people saw organic food being offered in the supermarkets they winced at the higher cost but took a few apples or carrots home with them.

But they were still relatively few. Organic food had yet to break out of its niche market. It was growing, but it was still only for the more eccentric customers. It was hardly a subject that dominated the board meetings at Tesco or Sainsbury's. Their share price was not exactly rocked by how many organic carrots they'd sold that month. And then, one March afternoon in 1996, the Secretary of State for Health, Stephen Dorrell, stood up in the House of Commons to make a statement which would shatter the nation's confidence in our food. This was to prove the turning point. The quiet revolution was about to explode.

Four months earlier Mr Dorrell had been asked on television whether there was any connection between BSE and a new variant of an old disease, Creutzfeldt–Jakob disease or CJD. He had said no, there was no conceivable risk from eating

British beef. Almost five years later Mr Dorrell was to tell me on the *Today* programme that he regretted ever having given that interview because he'd got it wrong. The message he delivered on that spring afternoon was a very different one indeed.

He told a packed and worried House of Commons that the 'most likely explanation' for the cases of CJD was indeed a link to BSE. MPs had been expecting bad news, but this was devastating. The implication was immediately obvious to everyone who heard him. Not only was mad cow disease wreaking havoc in the nation's dairy herds; there was a human equivalent of the disease and it was even more terrible. Every time we had eaten British beef in the past ten years or more we had been putting our health at risk. There was no cure for the disease and it was always fatal. CJD could lie dormant in our unsuspecting bodies for a very long time. It might take ten, fifteen, twenty years or even longer to make its malignant presence felt. No one could predict how many people were already infected. It might be a handful. It might be a million.

Even now, all these years later, it is hard to forget the impact of that statement. Many of us remembered with a shudder the steaks we had grilled or the Sunday joints we had enjoyed. Far worse, we looked at our children and thought of the mincemeat we had made into sauces for their spaghetti or the hamburgers we had treated them to on a Saturday afternoon shopping expedition. Could we really have been betraying their trust, poisoning our own children, condemning them to die from some hideous disease that we had never even known existed? It was scarcely credible.

The initial reaction was anger. We cursed the farmers who had forced their cows to behave like cannibals, eating the flesh of their own species. Then we cursed the arrogance and greed of the feed manufacturers and the renderers who had turned nature on its head by producing the wretched feed and not even telling the farmers what was in it unless they asked. Then we cursed the politicians and their civil servants who had been either uncaring or incompetent or secretive or all three.

It was at that point that many people decided they could no longer trust the food they were buying. Connections were made between cost and quality. There was one reason and one reason only for feeding ruminants on the ground-up brains and spines and bones of other animals. It was profit. The left-over bits of slaughtered animals cost the manufacturers relatively little and contained lots of protein. A cow fed on only her natural diet of grass and cereals produced nothing like as much milk as one fed on the concentrated food made from the left-overs from an abattoir. The manufacturers made more money and the farmers sold more milk and the price of food kept falling. So everyone was happy. Cheap food was what we wanted and cheap food was what we got. Year after year we had fallen for the notion that we should rejoice because we were paying so much less for our food than earlier generations and there was more of it. Plentiful and cheap: that was the mantra we were all required to chant. And so we did – until that spring day in 1996 when Stephen Dorrell dropped his bombshell on the floor of the House of Commons. It was then that millions of people began to appreciate for the first time the hidden costs of so-called 'cheap' food.

Since the middle of the 1980s we had watched those disturbing pictures on our television screens of cows slipping and sliding drunkenly across the floors of their farmyards in some macabre dance, their legs spread-eagled for all the world as though they were drunk. Now we were to watch their destruction. Not only the obviously sick cows but millions of their apparently healthy sisters were to be incinerated. The photographs of their grotesque funeral pyres began appearing in the newspapers, countless carcasses dumped one on top of another, their legs jutting out of the smoke and flames like some hideous painting by Hieronymus Bosch. In the years that followed 170,000 cows were to become infected and 4.7 million were to be slaughtered. British beef was banned around the world. No meat could be sold from cows that were more than thirty months old. We began to count the cost in taxpayers' money of compensating the farmers and paying the meat renderers. As I write these words the total is approaching five billion pounds and it is not over yet. Replacing surgical instruments that might possibly carry the vCJD infection will add hugely to the bill.

But worse – infinitely worse than all that – we began to share the grief of parents whose children fell victim to the hideous disease. And we began to worry for our own. Pictures of bright young teenagers began appearing in the newspapers and the text always told the same tragic story. Our beloved daughter had been full of life, full of fun, full of hope for the future. Then strange things started to happen. She could no longer concentrate. She grew irritable and angry. She became a different person. She lost control of her movements and her bodily functions. She became blind. She could no longer hear or speak. And then – with awful and absolute inevitability – she died.

Anthony Bowen is a bright little red-haired four-year-old. His mother died of variant CJD. Anthony was born three weeks before her death. It is known that a pregnant cow can pass BSE to her unborn calf. It is not known for sure whether a mother can pass CJD to her unborn child, but it is believed to be possible. It is hard to imagine the anguish of Anthony's father. He had to watch his wife dying from the disease. Every time he looks at his young son he must wonder whether he will die the same way.

The family of one young victim sent a video to Tony Blair. Donnamarie McGivern was a beautiful young woman, sixteen years old. She had been her school's athletics champion and a talented dancer. The disease hit her in the late nineties and the video showed what it did to her. She lay motionless in bed, her mother and sister stroking her hair. For almost a year she could not talk, eat or see. Then she died in the arms of her mother.

At the time of writing the number of people who have died like Donnamarie and Anthony's mother is still relatively low. By the time you read these words it may be much higher and heading higher still. It is one of the most chilling aspects of this dreadful disease that we can make no realistic predictions. In the autumn of 2000 a seventy-four-year-old man was confirmed to have been one of the victims. Until then, the youngest was aged fifty-four. How many other elderly people who were believed to have died from diseases such as Alzheimer's had, in fact, died from CJD? And in that same week a fourteen-year-old girl died. She had been ill since she was twelve. Did that mean that contaminated meat was getting into the human food chain in spite of the controls that were applied in 1989? We cannot answer any of these questions.

As well as being sent to Tony Blair, the video of Donnamarie was sent to Lord Phillips, the judge who chaired an inquiry into BSE that lasted more than two and a half years and cost an extraordinary twenty-seven million pounds. When Phillips finally produced his massive report there was one theme that ran right through it. The public had been misled about the risks. The people in power had adopted a policy 'whose object was sedation'.

As things turned out, the failure of that policy could scarcely have been greater. If the many false reassurances served any purpose at all it was the opposite from what had been intended. People began asking questions about what else might be going on to make our food less safe than we had believed it to be. If 'they' could do this to cows, what else might be happening on our farms and in our food factories? Of course it was possible that the tragedy of BSE was a one-off event, so horrifying that nothing like it would ever happen again because lessons had now been learned. On the other hand, it was equally possible that this was the inevitable consequence of messing about with nature with only the pursuit of productivity and profit in mind, treating animals with total contempt, ignoring the possible consequences. There was already a growing revulsion on the part of many people at treating animals as if they were nothing more than machines to be exploited and worked to death for our benefit. The lesson of BSE was that there were limits – not only moral, but practical. It might actually be in our own selfish interests to treat animals with a little more respect.

The other lesson was that in future we might have the courage to say no to a new technology before, rather than after, it has caused terrible damage. When I presented a special programme for *Panorama* on the BSE crisis Stephen Dorrell was desperate to assure me that the government had acted as soon as it 'had the evidence'. Who could quarrel with that? Well, the Livestock Standards Committee of the Soil Association had warned in 1983 that there were serious risks in feeding animal protein to ruminants and banned it for all organic farmers. They had no firm evidence, but every ounce of their intuition and common sense told them it was a stupid thing to do. They were ignored. The problem with waiting for the evidence is that by the time you get it, it may be too late.

The politicians, advised by their experts and their civil servants, also wanted to reassure us that every step is now being taken to ensure that we will run no more similar risks. When a few questions were raised about the safety of beef on the bone, even from animals deemed to be entirely free from risk, butchers were immediately banned from selling it and T-bone steaks disappeared from restaurant menus. But still doubts remain in the minds of many people. The most persistent is this: BSE and other, lesser, food scares such as salmonella in eggs were the result of an approach to farming and food production that placed quantity above quality at every step in the chain and yet that approach has never been seriously questioned.

If a chicken could be forced to grow faster to knock a few pence off the cost of producing it, we did it. And never mind that it lived in disgusting conditions that invited disease. If a cow could be forced to produce twice as much milk by feeding it ground-up meat and bone meal, we did it. If an acre of field could be forced to produce an extra ton of wheat by plastering it with synthetic fertilisers, we did it. And never mind where the chemicals ended up. If a field of carrots could be forced to deliver a bigger yield by spraying the plants with pesticides and the weeds with herbicides, we did it. And never mind that the Department of Health had to warn the people who ate those carrots to peel them carefully and throw away the top inch because of pesticide residues.

As so often before, many ordinary people with no great knowledge and no 'expert' advice to rely on were asking questions for themselves and reaching conclusions for themselves. One of those conclusions was that many of them wanted their food produced differently. They wanted less factory farming. They wanted food produced by less intensive methods. The supermarkets got the message. The revolution was born.

We are endlessly told – usually by the supermarkets themselves – that we have the best food retailers in the world and we should be grateful for what they have done for us. Not everyone is persuaded. Some of us regret the disappearance of so many small butchers and bakers and greengrocers and chemists. Some of us mourn the destruction of so many high streets and town centres. When the supermarket chains tell us they are not to blame we do not believe them. We prefer the evidence of our own eyes and the application of our own common sense. But in at least one respect the supermarkets cannot be faulted.

They are, by and large, run by brilliant marketing people who can spot a selling opportunity at a thousand paces and exploit it in less time than it takes to fill a trolley.

It took them a very short time to recognise that their customers were scared at the thought of genetically modified food and they dumped it from their shelves before the biotech companies and the politicians knew what had hit them. Nothing sounds louder in the ear of a supermarket executive than a customer's threat to take her business to a competitor. Now the sound of customers demanding organic food was becoming deafening. All those supermarkets which had so recently scorned the idea of the so-called beard-and-sandal brigade were suddenly fighting each other for every organic carrot and tomato they could lay their hands on.

Marks & Spencer had once loftily declared that it would not be bothering with organic food because there was not enough demand and the quality could not be guaranteed. When they finally realised what was happening they changed their minds in a hurry. Sainsbury's and Waitrose were stocking close to a thousand lines by the middle of 2000 and Tesco was catching up fast. Asda launched its own brand of organic food. By the middle of the nineties sales of organic food were growing at a rate of more than forty per cent year on year. In 1999 we bought about £550 million worth of the stuff.

At this point you might expect the men and women who run the organic movement in this country to be jumping up and down with delight and – for the large part – they are. But there are problems. One serious concern is how the supermarkets will behave in the next few years. So long as organic produce accounted for no more than a tiny niche market none of the big chains had been terribly interested. They sold the food at higher prices but paid the producers more for it, so their profits on the whole enterprise were relatively modest. They didn't worry too much about that. The main reason for stocking the stuff in the first place was to make sure that customers who wanted it did not go elsewhere and then buy all their other groceries in a competitors' store. But now things are different. One in three of us buys organic food at some time or other. Tesco sold more than £150 million worth in 2000. Across the whole market it accounts for nearly five per cent of total spending. Organic food has become seriously big business and every supermarket chain wants more of it.

The obvious way to take a bigger share of the market is to cut prices and then, maybe, push them back up again later when no one's looking. Tesco, for instance, said they would sell many of their organic products at the same price as conventional food. On the face of it, that sounds like good news; organic food has always commanded a pretty hefty premium which many people cannot afford. But the reason it costs more is that it is more expensive to produce. A supermarket boss might say: 'Oh, we wouldn't dream of forcing our suppliers to lower their standards.' They might even mean it. But if the biggest supermarkets run true to form they will do what they have done throughout their long and mostly profitable histories. Sooner or later they will put the squeeze on their organic suppliers just as they have put the squeeze on every other supplier over the years to sell to them at the lowest possible price. The organic farmers and growers might then decide that the only way they can survive is to cut corners and lower their standards.

All of this assumes that the extraordinary increase in the sales of organic food will continue. The projections suggest that it will. By 2002 it is estimated that it will have passed the billion-pound mark and, in the following five years, will have doubled again – at least.

Yet there is another little worm gnawing away at the core of what seems to be an exceedingly healthy, growing apple. There were two big reasons for the sudden leap in demand for organic food: fear and distrust. We feared the diseases that could result from factory farming and we distrusted – with good reason – the reassurances that we were endlessly given. Many people also fear the consequences of eating food that contains the residues of dangerous pesticides. Those fears do not apply to organic food. Organic crops are not sprayed with synthetic poisons, so there are virtually no residues. It remains the case that of the 170,000 cows infected with BSE only a tiny number came from organic herds and they were animals that had been brought into the herd from other farms. Not one cow born and reared in a fully organic herd has ever gone down with BSE.

Fear may be one reason for buying organic food. Moral objections to factory farming may be another. Concern for the environment may be another. Worries about the reckless use of antibiotics may be yet another. But there's something else too. Is organic food better for us? Surveys of public opinion and increasing sales in the supermarkets suggest that most people think it is. Hard scientific proof is less easy to come by.

Logic and, indeed, common sense suggests that it ought to be. If the soil is rich in micro-organisms and minerals then the crops that grow in it should benefit. And there is evidence that soil farmed organically is richer soil. The most recent serious study – in Switzerland – compared organic and conventional farming systems over twenty-one years. It showed dramatic increases in soil fertility. The mass of micro-organisms that deliver nutrients to the crops was up to eighty-five per cent higher in fields farmed organically than in those farmed conventionally.

But why should it matter how the plant gets its nutrition, whether it takes it from the soil or the farmers spreads it onto the field from a sack? If you see crops growing on a farm immediately after the fields have had a big dose of synthetic fertilisers they will look extremely healthy. When I started farming conventionally I was always amazed at the effect on the grass in the spring after we'd been out on the tractor spinning little pellets of nitrogen across the fields. You could almost see the grass shooting up with the first shower of rain. In no time at all it would be rich and lush. Fields that were too steep to drive a tractor on looked sickly by comparison. The effect of chemical fertilisers seems almost too good to be true. That's because it is.

There have been hundreds of studies in the last few years comparing the nutritional value of organic and conventionally grown food. Some of them have shown marginal benefits from eating organic food. In most cases the results have been inconclusive. But there is a problem with the way most of those studies were carried out. What many researchers have done is taken two plots of land, plastered one with manure and another with synthetic fertilisers from a bag, grown the crops and then analysed them. It makes a nonsense of the whole thing. You might as well compare the effect of being run over by a bus and being run over by a bicycle. The reality is that food cannot be sold as organic unless the land has been farmed organically for at least two years. It takes at least that long for it to recover its fertility and for the soil structure to begin its recovery. Only in the third year does it meet the standards required. So the only valid comparison is between crops grown on land farmed and certified as organic and crops grown conventionally.

When all the flawed studies have been eliminated the picture becomes clearer. Organic food contains more of the things we want – such as dry matter, minerals, trace elements and Vitamin C – and fewer of the nasty things such as nitrates and heavy metals and pesticide residues. Another benefit is that it contains more secondary nutrients or 'secondary metabolites'. These are the plant's own defences against pests. Over the years we have learned that they are good for humans too. They include antioxidants such as flavenoids and phenolic compounds which help boost the immune system.

Critics of organic farming are not persuaded by any of this. The thoughtful sceptics say there is not enough evidence. They have a point; much more research needs to be done. But there is a powerful lobby opposed to organic farming who have no interest in evidence or research. That lobby includes the vested interests: the agrochemical companies who sell the pesticides and the synthetic fertilisers; the biotech companies who want to sell their genetically modified seeds; the barley barons who have made small fortunes farming for subsidies; the politicians who are afraid to admit that they may have got it wrong over the years and are afraid to upset the big vested interests. There is more than half a century of evidence and a mountain of research to show that the old system has let us down. But they have only one concern: to defend their own interests by defending the status quo.

When I bought my farm in West Wales all those years ago I could scarcely have been more naïve. I barely knew the difference between a ton of silage and a farm gate. At least I had the sense to hire a manager, Eddie Cooper, who had plenty of knowledge and even more enthusiasm. He needed it. I was still a foreign correspondent in those days and I happened to be travelling between Peking and Hong Kong on the day we took possession of the farm. I phoned Eddie that night from my air-conditioned hotel room overlooking the splendour of Hong Kong harbour. It was hard to imagine being back in a rain-sodden, windswept, rundown dairy farm on a Welsh hillside, but I tried.

'How's it going?' I asked.

Eddie was always the master of understatement.

'Oh, fine,' he said. 'Apart from the fact that the bugger who sold you the farm has flogged off every last bale of hay and pound of silage and since it's still winter there's not a blade of grass in the fields. So we've got sixty-five starving cows and nothing for them to eat. Apart from that – no problems.'

Eddie coped. Thank God for a good manager and kind neighbours.

I was never much of a farmer. I suspect good farmers are born and not made and I don't suppose I ever will be much good at it. But I knew what I did not like and I did not like spreading chemicals all over the fields and stuffing the cows with antibiotics. So when Patrick Holden and Peter Segger appeared on the scene I was ready to listen.

They were farming in the next valley and their farms were organic. Both of them have gone on to great things since then. Patrick runs the Soil Association and is one of the most influential voices in agriculture these days. Not that he's happy with what he's achieved. He thinks it is possible to turn the Soil Association into a mass member organisation of people who recognise the connection between good farming and good food. Peter built up the most successful organic food distribution businesses in Europe. They are both remarkable men and are widely respected not only in this country but around the world. But in those days they were regarded – at least in my part of Wales – as bonkers. The idea of farming without synthetic chemicals was seen as so eccentric they might as well have been trying to grow pineapples or passion fruit on the summit of Mount Snowdon. So when I announced that I was going to follow their lead and convert my farm to organic the locals in my area shook their heads and tried to hide their smiles.

They would never say it to my face – they were far too polite – but I could imagine the conversations in the local on Saturday nights.

'Silly bugger. How's he going to grow enough grass for all those cows without nitrogen, eh? Ah well, he'll learn. Must have been crazy in the first place, coming down here from London and buying that farm the way he did, knowing nothing. Still, he shouldn't listen to those crazy blokes from Lampeter. Nothing more than hippies, really. Led him astray. Shame isn't it?'

It was not only in West Wales that farmers thought people like Patrick and Peter were crazy. There were no more than a handful of people farming organically and organisations such as the mighty National Farmers' Union seemed to take no notice of them at all. That has changed. In 2000 so many farmers applied for grants to convert to organic operations that the scheme ran out of money in the first few months. When Patrick took over the Soil Association it had a dozen staff, mostly part-timers. Now it has closer to two hundred and can scarcely cope with supplying information and inspecting farms for their organic certification.

It would be good to think that all this is happening because the likes of Patrick and Peter have won the argument against factory farming but it's not that simple. The industry has been in crisis in Britain for many years now and the only way to make money is to sell at a premium price. Some organic dairy farmers, for instance, get almost twice as much for their milk as conventional farmers. So most of those who are converting to organic operations are doing it because it offers them an economic lifeline. It's that or bankruptcy. I know one farmer in Wiltshire who has a herd of three hundred cows. When I first talked to him about organic farming in the late 1980s he treated me as though he were a mathematics Nobel Prize-winner dealing with a five-year-old learning his twice-times tables. Ten years later he began the process of conversion, cursing that he had not done so many years earlier.

Even so, he is still in a tiny minority. Only about three per cent of the arable land in Britain is farmed organically. The real arguments have yet to be won. Perhaps they never will be. But at least there is one great myth that should be laid to rest. It says that to deny the superiority of intensive agriculture is to deny science. For most of the twentieth century its propaganda went virtually unchallenged.

It is a potent myth because it sets up a straw man and then proceeds to demolish it. Over and over again it asks this question: how can anyone be taken seriously who does not believe in science? On the face of it, that is a perfectly sensible question. The flaw in it is that no one is denying science. What the critics of industrial agriculture are asking is something quite different: 'Why should we continue to have faith in a system that is based on too many flawed principles?'

Real scientists know – and are happy to acknowledge – that the essence of science is fallibility and that all the knowledge they claim is provisional. It is not scientific knowledge unless it is in a form that leaves it open to being disproved. That is the principle of falsification laid down by Karl Popper. As Jonathan Dawid, a physicist at Harvard University, puts it: 'Every scientist is aware of the limitations of his craft. How can it be otherwise in a profession where each step of progress requires the consignment of another man's work (and often one's own) to the dustbin?' I think of those who deny that basic principle as blind worshippers. Blind worshippers believe in infallibility. It is the fig leaf with which they try to conceal their ignorance.

Those sad souls want an authority in which they can trust implicitly. They seek to invest scientists with powers denied to mere 'laymen'. There is nothing new in this. Human beings down the ages have sought infallible authority; it's so much more comforting. The blind worshippers who might once have got it from priests often now look to scientists to fill that role. It is easy to understand why. The world is such a complicated place that it's difficult to work everything out for ourselves and we need someone to tell us what the truth is. Scientists are the people who know the most about how the world works so they are the authority to whom the blind worshippers turn. Good scientists endlessly remind us of their fallibility but the blind worshippers refuse to listen; doubting priests were always a bit of a pain.

The trouble is that there are always some scientists who are perfectly happy to be worshipped. Their white coats are their vestments; their hierarchy of degrees and academic titles represent the order of priesthood. To the great unwashed their arcane language is as mystifying as any Latin liturgy. They disappear into the vestry of their laboratory with its mysterious equipment and when they emerge they smile beatifically as we bend the knee to receive the blessing of their great knowledge. If their doctrine contains any notion of fallibility or the provisional nature of their knowledge it is written in very small print indeed. These are the false prophets. As the priesthood has its charlatans, so the profession of science has its own pretenders, though there are mercifully few of them.

Science is simply knowledge acquired by study. It is nothing more than that and nothing less. It is a powerful force and it is open to every single one of us with an enquiring mind to engage in it. It helps to have training. It helps even more to have research support and resources. But none of that is essential. Knowledge may be acquired by sitting in a laboratory and looking through a microscope and learning from it. Equally it may be acquired by sitting on a farm gate, studying the way grass grows and learning from that. It may be the ability to read the map of the human genome and explain it. It may be the ability to read the way a pig behaves with her litter and explain why. It is not the ability to experiment with different molecules and create a powerful pesticide any more than it is the ability to experiment with different rotations of crops to create a bigger harvest.

We say that a man is a scientist if he has studied at university and is now engaged in transferring genes from one species to another to create a superior plant. But we do not say the same of a man who has studied plants as a humble gardener all his life and learned to cross-breed from one variety of plant to another to achieve similar results. He is still just a gardener. The real difference is that one will certainly have letters after his name and the word 'Doctor' in front of it. The other will not.

The blind worshippers tell us that without 'science' as they narrowly define the word there would be no progress. Yet modern agriculture began more than ten thousand years ago. How many laboratories were there then? The answer is, of course, that there were thousands upon thousands of them. Every enclosed and cultivated patch of ground was a laboratory. Every seed sown was an experiment. If the first primitive man who spotted the effect of animal dung on the growth of his plants and managed to repeat the effect was not a scientist, then what was he? He acquired knowledge. He discovered fertiliser. He proved that a plant was able to thrive with it and would fail without it.

He may not have known why it worked but his descendants found out and learned how to create compost and apply that to their crops. They acquired knowledge. They learned that certain insects were beneficial and should be encouraged and others were not and should be deterred and they found ways of keeping them under some sort of control. In what sense were they different from a man in a white coat who sits in his laboratory and manipulates molecules to create poisons to kill insects?

The blind worshippers will say most of these things came about by accident or by trial and error. I agree. So have many, if not most, of the great 'scientific' discoveries of our time. Remember Alexander Fleming? The difference was that Fleming was able to write up the results of his experiments and repeat them under laboratory conditions, albeit primitive conditions by today's standards. Those prehistoric farmers were not. Did that make their discoveries less valid?

And what of those 'real' scientists who get it wrong? As every researcher knows, it happens all the time. I spoke to one of Britain's most distinguished biologists in the course of researching this book and he confessed to me that he had been teaching his students something for thirty years that turned out to be wrong. It had been scientific dogma – unquestioned in a thousand learned papers – until it was overturned. But the blind worshippers at the altar of science cannot accept the doctrine of scientific fallibility. It might undermine their belief. That's the trouble with blind worship. You fail to spot things when they start going wrong.

The blind worshippers failed to spot some of the most catastrophic mistakes in the history of agriculture. They failed to spot what might happen if we turned cows into carnivores. They failed to spot the possible consequences of hundreds of scientists happily creating antibiotics to be injected in animals as though they were Smarties fed to children when they should have known what devastating effects their creations would have. They were happy to spray our fields and hedgerows with persistent toxins that poisoned the land for generations, killed the wildlife and endangered our health. And they did not warn us what might happen to the soil if we persisted with practices that destroyed its population of microorganisms.

I know many people who worried about all those developments and argued fiercely against them. Some devoted their lives to building persuasive cases. But for the most part they were ignored. Why? Because they were not scientists. They had no letters after their names. Their work had not been 'peer-reviewed' by other scientists and published in any distinguished journals. But they read the work of other scientists and compared it with their own experience. They acquired a great deal of knowledge of their own through observation and practice. So what in God's name are they?

I'll tell you what the blind worshippers say. They are cranks. Worse, they are cranks engaged in 'bogus science'. It's one of the oldest tricks in the book. You pin a label on someone, backed by the authority of your many credentials, and sit back with an amused smile as he struggles to detach it. Let me introduce you to two of the cranks it has been my privilege to encounter over the years. Both of them had come to the conclusion that there were serious problems with some modern farming methods and there might be better alternatives.

One of them is Richard Young of Kite's Nest Farm in Worcestershire. The farm is owned by his mother, Mary, and he runs it with his sister Rosamund. It lies on the northern escarpment of the Cotswold Hills, most of which are covered these days in regimented rows of cereals marching towards the horizon. In many respects Kite's Nest is a time warp. It is a mixed farm, as were most half a century ago before subsidies dictated farming methods.

The cows have names rather than numbers. They live in families and suckle their youngsters. The calves are never removed forcibly from their mothers. When her milk dries up the calf is weaned away naturally and when the new calf comes he moves aside. But he will stay with his mother. There might be several generations in one family. They graze in pastures filled with the sorts of plants that would have been killed off long ago if they had been ploughed and plastered with synthetic nitrogen. They are rich in herbs, which the cows select carefully. There is wild thyme and clover and bird's-foot trefoil. The cows need no scientists to tell them what is good for them and where they should go to get the nutrition they need. They rely on instinct developed over millennia.

The farm is a botanist's dream. There are rare plants, including the yellow-flowered Dyer's greenweed, pyramid orchids and clustered bellflower and a hundred other species that have long since vanished from most of Britain. In the woodland glades there are harebells and cowslips. You do not see all this on a modern efficient farm.

But who is to say that this is not efficient? The cows live longer and are healthier because they are under no stress. When you approach them they do not back away. They have been treated with care and affection all their lives and have no fear of people. But they must earn their keep and when their youngsters reach the right age they are slaughtered locally. The Youngs sell the meat in the farm shop. Because of the way the animals have been reared it fetches a premium price and there is no shortage of buyers; the customers know they can trust this food.

When other farmers were forced to slaughter their precious livestock and watch their life's work being destroyed by BSE, the Youngs had no fears. Mrs Young had stopped buying compound feed in 1974 when she discovered that it contained chicken carcasses. She did not ask the men in laboratories if it was wise to feed a grass-eating animal on the flesh and bones of other animals. Just as well. If she had, they'd have said: go right ahead. As it is, the Youngs have never had a case of BSE. Not one.

So this is a thoroughly old-fashioned farm. It survives because its costs are low and it caters to a specific market. It is well managed and it is beautiful. On a spring day when the flowers are at their best and the insects are stirring and the woods are bursting into leaf there is a real sense of nature in harmony with itself. The blind worshippers will say that this is romantic twaddle. There is, they will say, no such thing as 'nature'. There are trees and there is grass; there are insects and there are animals. But they do not add up to this mystical thing romantics call 'nature' and which they invest with all manner of meaning. The blind worshippers will tell you that nature is to be controlled, managed, manipulated. If we relied on farms like this we would all starve. It follows, therefore, that Richard Young is a crank. Well, let me invite you to leave his fields and spend an hour in his office.

It is a big office, the first floor of an old barn. This is another realm, far removed from the world of ancient meadows and cows with names. Computer screens glow, fax machines stutter out reams of paper and the filing cabinets which line every wall bulge with thousands of documents. Few of them have anything to do with the farm. Most are scientific papers, the sort that have been published by clever men and women with enough degrees to paper their walls. They are overwhelmingly concerned with the world of microbiology and, more specifically, the effect on bacteria of antibiotics.

Richard Young may not have a single academic letter to his name but he spotted the devastating assault being made on the front line of our medical defences long before many of the people who subsequently wrote these papers. For fifteen years he has been assembling a powerful argument and trying to get people to listen. Naturally he is not qualified to stare into a microscope and write up the results in a way that will be approved by the blind worshippers. The only qualifications he possesses are an enquiring mind, an ability to absorb, understand and analyse a great deal of information, an intimidating capacity for hard work and a lifetime of studying the way animals respond when they are treated in certain ways. But he is not a scientist. I leave you to judge whether his concerns over antibiotic resistance were well founded or whether he is just a crank.

Mark Purdey is another organic farmer and, until recently, dismissed as just another crank. He is the perfect object of disdain for the blind worshippers. Back in the eighties he looked at the 'scientific' solutions being offered by the Ministry of Agriculture and did not like what he saw. He refused to obey official orders to dose his cows with high levels of an organophosphate poison. All the scientists at the Ministry knew it made sense. It was the only effective way to deal with a pest known as warble fly, they said. But Purdey thought it was crazy to pour a poison derived from a military nerve gas along the spine of every cow so that it would seep through the skin and turn the animal's insides into a hostile environment in which no parasites could survive. The Ministry took him to court to make him do it. He still refused and the case ended up in the High Court.

On one side of the courtroom was a government ministry, a powerful chemical manufacturing company and the might of the scientific establishment, all with the same message: OPs were harmless to man and beast, even at the high levels recommended, and only a crank would argue otherwise. On the other side was Mark Purdey, a Somerset farmer, self-educated and stubborn. Purdey won.

The phone started ringing at the Purdey farmhouse even before he'd boarded the train to return home: one farmer after another, all of them telling him the same thing. You're right, they said, I've been using OPs just as the Ministry ordered for years now and I'm suffering because of it. Their symptoms were remarkably similar. All of them were consistent with some kind of disease of the nervous system. For years they had been told they were imagining it. Now they had a champion who had proved to the satisfaction of one of the highest courts in the land what they had always known: OPs were destroying their health. But it was the effect of OPs on cattle that Purdey became particularly concerned with.

In the 1980s he had observed a link between BSE and the use of one particular organophosphate. Nobody listened to him. Of course not; he was a crank. He tried to get some money for research from the Ministry but was turned down. There were two hundred different projects being funded but Purdey was not a real scientist so why should he be indulged? Even so, he kept working away at his theory. He used his own money to embark on the most extraordinary detective hunt, looking for clues not just in this country but around the world.

The first thing he found was that there was a particularly high incidence of BSE in those areas where the warble fly-killing OP was most intensively used under orders from the Ministry. He also found that there was no record of cows dying from BSE that had been born and raised on fully converted organic farms where virtually none of the chemical was in use. Then he began finding similarities between the symptoms of BSE and chronic low dose OP poisoning. When he started looking abroad he discovered another link. It provided evidence for his theory that one of the effects of this OP was to reduce copper levels in cattle. If that was simultaneously compounded by an excess of manganese, he believed, it could lead to BSE.

So he looked for clusters of BSE-type diseases around the world. He found them in areas where there were abnormally high levels of manganese in the soil and vegetation and low levels of copper. He also found that human versions of BSE were more likely to afflict people where there were high levels of manganese. In Chile, where manganese oxide is mined, the mine workers are affected by what is known locally as 'manganese madness'. It is, in many respects, remarkably similar to variant CJD.

All fascinating stuff but it has been decided officially that contaminated animal feed is the single reason for BSE and that's that. Well, maybe, but you will notice I say it 'has been decided' as opposed to 'it has been proved'. That's because it has not been proved. Contaminated food is, in the minds of the scientists, the most likely explanation. They have been unable to prove it for sure, still less to prove that there may not have been other factors. And anyway, Mark Purdey is not a real scientist. True, he has learned a great deal about the brain's complex biochemical pathways but he has no degrees. He is self-taught and, therefore, easy meat for the blind worshippers. David Brown is more difficult for them to dismiss.

Dr Brown specialises in prion research at the Department of Biochemistry at Cambridge University. He read the Purdey findings and tried some experiments. He has now shown that when manganese is added to cell cultures in the absence of copper it bonds to normal protein to create the 'rogue' prion protein that is found in victims of BSE and variant CJD. It was an American scientist, Stanley Prusiner, who first suggested that a 'rogue' prion kills animals and humans by converting normal prion protein molecules into dangerous ones that cause degenerative diseases of the brain. That is now widely accepted in the scientific community. Subsequent research has shown that there is a tenfold increase in the level of manganese – as well as low copper – in the brains of all those who have died of vCJD.

As Purdey points out, BSE arose during the 1980s when cattle farmers were forced to pour a specific OP along the spinal column of their animals. The doses demanded were higher than anywhere else in the world. The effect of this OP is to deplete copper as well as to transform manganese into a lethal 'free radical'-generating species in the brain. At the same time, cattle feed was being supplemented with chicken manure from birds dosed with manganese to increase their egg yield. So the prion proteins in the cows' brains were both deprived of copper and dosed with manganese.

But let us dismiss both Mark Purdey and his scientific supporter David Brown and turn briefly to another 'real' scientist. Professor Alan Ebringer is an immunologist at King's College, London. He believes that BSE and variant CJD are the result of an immune reaction that has gone out of control. He believes that, with the help of another professor at King's, he has identified the culprit. It is a common microbe found in the soil, sewage and water supplies. It is the same microbe that is responsible, says the professor, for multiple sclerosis. He thinks variant CJD is an extreme manifestation of MS.

So there you have another factor that may have contributed to BSE.

Now all of this presents serious problems for our blind worshippers, those who treat 'bogus' science with such scorn and say we should pay heed only to 'real' scientists. Who are the real ones and who the bogus out of the cast of characters I have just introduced?

Surely Richard Young must be bogus; he has not a single scientific qualification to his name. And yet the theories he and the Soil Association have argued so fiercely on antibiotic resistance are now shared by every serious scientist in the field. The documents he and his researchers have produced over the years have made a significant contribution to the debate. They have been vital in raising its profile so that politicians have at last begun to pay serious heed to what is going on.

Well then, how about Mark Purdey, another farmer with the temerity to tell the scientists that they'd got something badly wrong? Fine, let us consign him to the stocks of scientific ignorance as well and beat him about the head with rolled-up copies of learned publications as we mock his outrageous theories. The problem with him is that we shall have to mock one or two rather serious scientists as well, such as the researchers who proved in the laboratory so much of what Purdey researched in the field. There is also the inconvenient fact that, after years of being ridiculed, vilified and persecuted by Ministry officials and the scientific establishment, he has now had his findings published in serious scientific journals and has been given research money to carry on his work.

So let us at least condemn as bogus those who hold that a common bug is to blame for BSE. But that might be tricky. They are both distinguished professors, men of great scientific knowledge. How can they be bogus?

The answer is that we can dismiss none of them. They have all, in their different ways, pursued knowledge and tried to prove something. They are all worthy of our attention. If the blind worshippers cannot see that, then it is they who should be dismissed.

Science has long passed the stage where an expert can be expected to know all that there is to know in his discipline, let alone in other related disciplines that might be affected by his work. We have simply learned too much. That means an increasing amount of specialisation or compartmentalisation and that creates problems. A scientist who specialises in the molecular biology of plants and has learned how to transfer a gene from another species may know little or nothing about the effect of that plant being eaten by an animal or a human. Equally, the scientist who knows about nutrition and the way the human body works may have no real knowledge of what makes a transgenic organism tick.

They might each say: why worry? Nothing will be approved for release until it has been authorised by the regulatory authorities. And that's fine if the men and women who sit on the relevant committees know all that they need to know. But they do not. Most of them are not scientists in the first place. Even those who are scientifically qualified need the most up-to-date information. That is difficult in a scientific area such as this, where new problems and new discoveries arise every other day. It takes a long time for new scientific information to be published. By the time it reaches the relevant committee it may be too late.

And there is another problem with new research: fraud. Three of the most prestigious medical journals in Britain are the *Lancet*, the *British Medical Journal* and *Gut*, which deals with intestinal matters. For a doctor or medical researcher to have a paper published in any of them is a high achievement. It cannot happen unless the work has been reviewed by other qualified experts working independently, a process called peer review. So, in theory, the work can be trusted. The reality can be very different.

In December 2000 something most unusual happened: the same letter appeared in all three journals. It was a deeply worrying letter about research fraud being perpetrated by doctors who want the money and the kudos that go with the publication of work in a leading journal. The letter was signed by the editor of each one of those journals.

It was not the first time that the editors – highly qualified and well-respected men in their own fields – had raised the question of fraud. A year earlier they had set up the Committee on Publication Ethics (COPE) as a forum in which to discuss research misconduct. In less than two years COPE dealt with 110 cases of fraud or ethical misconduct. They believe that is a fraction of the true figure, if only because many editors do not bring cases to COPE. As Professor Michael Farthing, the editor of *Gut*, said at the time: 'We have no idea still what fraction of the whole this represents.'

It is not the job of editors to stamp out research fraud and stop researchers faking their results. The responsibility for that lies with the General Medical Council and the various Royal Colleges. The editors wanted them to set up a national body to root out the fraudsters and deter others from trying the same. According to the editors, they promised to act and they did not – hence their return to the fray. 'The GMC and the Royal Colleges have not shown the resolve they said they would show,' said Dr Richard Horton of the *Lancet*.

Some of the fraud is frankly silly and easy to spot. Professor Farthing described the case of a researcher in the United States who used a black felt tip pen to colour in patches of transplanted skin in white mice. But some fraud can be very difficult or even impossible to spot at the time. A doctor at King's College Hospital in London was suspended by the GMC after carrying out a study investigating what damage might be caused to the small intestine by a particular drug. His research was based on the urine samples of twelve patients. It turned out that all the samples had been provided by the doctor himself. It took ten years to bring him to book. In the meantime his research was published in one of the most reputable journals in the land.

This is more than a family squabble between members of a notoriously sensitive profession. It matters to all of us. If the research is faulty or fraudulent how can we trust the judgements based on that research? They may be judgements that will determine whether a new drug is approved. Or a transgenic food. Or a new additive in our food. Or a new, toxic compound to be sprayed on growing crops.

There is another question worth asking: who pays for all this research in the first place? The answer, overwhelmingly, is industry. They have their own research departments – enormous ones in the case of the biggest biotech companies – but they also fund research projects in universities. The old days when much of the research in our great academic institutions was paid for out of the state's coffers, or by disinterested benefactors for the greater good of society, have long gone. The paymaster now is almost always a company with a specific enterprise in mind. Most of the results of the research are kept secret in the interests of commercial confidentiality.

Some research, of course, goes into safety but not much. A study by Dr David Barling and Rosemary Henderson at the Centre for Food Policy of Thames Valley University found that 'private sector research into safety factors is conducted to ensure regulatory approval for commercialisation'. So the amount spent depends strictly on the demands of the regulators. I have already described how limited are those demands. So who is paying for the more basic, fundamental research into the environmental and safety implications of GM foods or any other aspects of industrial agriculture? Well, some of the funding comes from the public sector but, once again, precious little.

Barling and Henderson praise the work that has been done but say it is simply not enough: 'The government needs to rethink the role of public sector science and technology research and focus on a revised perception of the public interest that prioritises safety and health issues.' They paint a picture of a 'hive of industry at work for UK plc' and conclude dryly that, as far as fundamental safety research is concerned, 'the private sector's research priorities lie elsewhere'.

Dr Mae-Wan Ho puts it more graphically. She distrusts what she calls the 'scientific establishment' because it is 'increasingly in bed with big business'. Our academic institutions, she says, have 'given up all pretence of being citadels of higher learning and disinterested enquiry into the nature of things, least of all being the guardians of public good'. Dr Arpad Pusztai, who worked at one of the most respected research institutes in the world until he publicly voiced his misgivings over the possible dangers of GM food, says that the industrial scientist is no longer a free agent. He is 'hired for a particular job which is restricted in scope and objectives and carried out under close supervision'. Pusztai says that 'data obtained becomes the property of the company that pays for the research. In most instances the scientist has no rights to discuss or publish the results without the permission of the company and the company may withhold publication for five years.'

He claims that even university or government research scientists are not in a much better situation because of the conditions to which they have to agree when they sign up to do the work. From his own experience he concludes: 'It seems that, in the eyes of many senior scientists today, the future of science lies with industry. When scientists who have no obvious financial connection with the biotech industry defend GM crops so blindly and attack even the mildest critics, slandering their work and abilities in the process, we must ask ourselves what motivates them. And one possible motivation is that, with the rapid disappearance of the State patronage of science, many of these people are genuinely worried about the future funding of scientific research.'

We have had one agricultural revolution in living memory. It began in the 1950s and its effects – for good and ill – are with us still. It gave us bigger harvests. It also gave us environmental destruction and pesticide residues and antibiotic resistance and the horrors of mad cow disease. The more we have learned about the food on our plate, the way it is grown and the way it is processed, the less we have come to trust it. Now we are being invited to embark on another revolution – genetic modification – with consequences we can only wonder at. I referred earlier in this chapter to the counter revolution that began in the closing decade of the twentieth century: the signs of a movement away from factory-farmed to organically grown food. But what we need is more than just a change in farming methods. We need a new approach.

I began this book by describing the events that led to the

development of intensive agriculture. With the benefit of two generations of experience it is easy to see why it went wrong. It was never more than a knee-jerk reaction to the fear of food shortages: produce it cheap and make sure there's lots of it. None of us paused to think about the long-term consequences of what we were doing. We must not make the same mistake again. We must think seriously about what we expect from food and the way it is produced. And our starting point should be that agriculture is, in effect, the nation's primary health service.

Good health is more than just the absence of disease. It is impossible to define it in terms that would satisfy the scientist or even the medical student but we know it when we see it. I wrote earlier about the sense of wellbeing we experience when we look at a healthy, happy baby. He lies there, with a full stomach and an empty nappy, gurgling and chuckling away for no particular reason and it just makes you feel better. He's full of life, bursting with some indefinable force. Sure, it might all change in the next hour and there will probably be tears before bedtime, but for those few moments there is something special. How to explain it? Of course it's partly because there is nothing wrong with him; he is free from disease and has experienced none of the trials of life that will eventually bow his shoulders and put lines on his forehead. Yet all our senses tell us that there is something more to it than that. We can feel it and we relish it. If we get it right the food that baby eats for the rest of his life will contribute to his health. Yes, we all know that if he eats enough fresh fruit and veg and not too much fatty meat or too many cream cakes, he is less likely to have an early heart attack. But that's just the absence of disease. Good food – grown in healthy soil and rich in vitamins and minerals – will promote his health. It will strengthen, not weaken, his immune system. If we deny him that we betray him.

So this is about more than the merits of one group of farmers over another and apportioning blame and praise. We should be moving on from that. We should be less arrogant and accept that good agriculture should not simply be battling against nature. There is always an element of that. If we allowed nature to have its way entirely most of us would starve. But it is possible to farm in harmony with nature and not always in conflict. We should also acknowledge that good food is about more than calories and vitamins.

You don't have to live in a yurt and wear rope sandals to accept all of that. Nor do you have to go back to Aristotle, who believed that every plant, let alone every animal, had a soul. It is possible both to accept the basic scientific principles of cause and effect and also to believe in the holistic view of the world as a living organism. What is impossible to believe in is the arrogant notion that we can do what we damned well like and to hell with the consequences. That's what we have been doing for the past half-century. It is time to stop.

IF I MAY JUST FINISH

Q&A

I have always felt sorry for authors who appear on programmes like my own to be interviewed about their new book. They've probably spent years researching and writing it and they're lucky to get three or four minutes to tell the audience what it's all about. And those few minutes are punctuated with questions from a grumpy interviewer who has almost certainly not read the book and is relying on a brief from an overworked researcher who has also not read the book. To add insult to injury, the interviewer is probably acting as devil's advocate because that way you're more likely to have an argument and arguments make the best listening. Or so we believe. Pity the wretched author.

By the time you read these words it may have happened to me. Since I have done more than my share of it in the past I can hardly expect any sympathy. So what follows is an attempt to get my retaliation in first. This is the transcript of an interview between the author of this book and a grumpy old interviewer early in the morning on Radio 4. I freely admit that it's a great deal easier to give the answers if you have also written the questions – and if you're not being interrupted all the time.

INTERVIEWER: You tell us the food we eat is killing us so how come we're living longer than ever before?

AUTHOR: I didn't say it's killing us –

INTERVIEWER: – sounds like that. All this stuff about the 'great food gamble' . . .

AUTHOR: That's a different point. All sorts of things used to kill us off early in our lives that don't now. We started living longer the moment the Victorians built some proper sewers and gave us clean water and sanitation. That put an end to plagues for a start. There are dozens of other factors. Even the motorways helped. The number of young men killed in accidents has fallen by half since the fifties. Infant mortality was nearly six times higher in 1948. In those days one woman in a thousand died in childbirth. That's fallen by about ninety-five per cent. Antibiotics have been a huge lifesaver: no more epidemics of polio and diphtheria and TB. We've found cures for all sorts of diseases, and surgeons can do things now they wouldn't have dreamed of a few years ago. If you want the really big reason we're living longer today it's better medicine –

INTERVIEWER: – and a better diet. Look how tall children are today compared with the scrawny kids when we were young. That's because there's more food grown as a result of what you disparagingly call industrial agriculture. And that means the food is less expensive so there's less malnutrition because people are eating more.

AUTHOR: Yes, we're eating more but are we eating better? Children may be taller but they're also a lot fatter. And that's causing all sorts of problems. Even the government admits that the life expectancy of many of today's children will be years shorter than their parents' and that's partly because of poor diet. Obesity has tripled since 1960. Asthma has doubled in the last ten years. One child in seven now has some sort of respiratory disease. What's so good about that?

INTERVIEWER: But you can't blame the farmers for the way kids eat.

AUTHOR: I'm not blaming the farmers for anything. I'm blaming the system –

INTERVIEWER: – and what the system has been doing is, entirely sensibly, encouraging farmers to grow more food more efficiently.

AUTHOR: Yes, and that's all it's been doing. No one's been looking at the other side of the balance sheet –

INTERVIEWER: – because there's no proof that it's doing us any harm. We're fitter than we've ever been.

AUTHOR: No, we're not.

INTERVIEWER: Yes, we are. You've already acknowledged that we're living longer so that proves it.

AUTHOR: It doesn't prove anything of the sort. I've tried to give you some of the reasons why we're living longer. And anyway there's no point in living longer if we're not fit enough to enjoy it. The Office of National Statistics has published some interesting figures on that. Its report acknowledged that there has been a dramatic increase in life expectancy over the last century but it asked whether all those extra years are lived in good health. Here's what it concluded: 'Healthy life expectancy did not increase by as much as total life expectancy, with the result that both men and women are living more years in poor health or with a limiting long-standing illness.' Puts a different complexion on it, doesn't it?

INTERVIEWER: Maybe. But you can't prove that has anything to do with the food we eat.

AUTHOR: No, you can't prove it. But if you look a bit more closely at all those statistics you come up with some interesting stuff. The things that used to make us ill were mostly infectious diseases. Ten times as many people died from them in 1948 compared with fifty years later. So what d'you think is the big disease now?

INTERVIEWER: You're going to tell me it's cancer.

AUTHOR: Indeed I am.

INTERVIEWER: But we've just had figures published showing some big falls. There's been a real drop in lung cancer cases for instance.

AUTHOR: True, but that's because so many people have stopped smoking. There have been huge increases in many other cancers. Overall it's gone up by sixty per cent since 1950.

INTERVIEWER: And you're saying it's because that was when intensive farming with chemicals began?

AUTHOR: No, but there has been a massive increase in the use of man-made chemicals. It's risen 600-fold since 1940. Nobody disputes that many of those chemicals can cause cancer.

INTERVIEWER: And most of those chemicals have got nothing to do with agriculture. They're in the air we breathe. They're everywhere.

AUTHOR: But an awful lot are used in agriculture . . . and some of them end up in the food we eat. The point I'm making is that it's ridiculous to be exposing ourselves to any more chemicals than we absolutely have to.

INTERVIEWER: But the reason more people get cancer may be simply because we're living longer.

AUTHOR: No, the statistics are adjusted for age. In fact, if you took chemicals out of the equation you could make a very strong case for why there should be less cancer, not more. As you suggested to me earlier, we eat a much more varied diet than we used to – far more fresh fruit and veg – and that ought to help protect us for a start.

INTERVIEWER: Okay, but we're winning the war against cancer. Hardly a week goes by without some new treatment being announced or some new discovery being made. They're spending a fortune finding a cure for cancer and they'll get there one day.

AUTHOR: Maybe, but don't hold your breath. And wouldn't it make sense to spend a bit more money trying to find out why so many people get cancer and trying to do something about it rather than trying to cure it after we've got it . . . and usually failing? The fact is, nearly forty per cent of the people in this country will get cancer. If they don't die *from* it they will die *with* it. That's appalling.

INTERVIEWER: So how do you explain all that?

AUTHOR: Well, old cynics like me always look for motives and financial incentives. There's a lot of money to be made from manufacturing chemicals. There's a lot of money to be made from selling drugs that cure cancer. There's no money to be made in expensive research that produces answers manufacturers may not like. We're back to our old friends, the vested interests. If big business is funding more and more scientific research it is doing so in order to make a profit. That's why virtually all the money spent on research in farming has gone towards intensive agriculture and factory farming. That's the sort of farming that needs to buy the products made by the companies doing the research. That's where the largest profits are to be made. How would it profit an agrochemical company to pay for research into small-scale, low-input farming? It wouldn't. So they don't do it.

INTERVIEWER: So you're another of those conspiracy theorists, eh? It's all the fault of filthy capitalism!

AUTHOR: No, I'm just a realist. I don't want to destroy the system; I just think that we should always be looking for ways to improve it. And we should at least recognise the influence of big business. Professor Michael Crawford is the director of the Institute of Brain Chemistry and Human Nutrition and he makes the point that every medical school has a full department of pharmacology, researching and teaching at postgraduate level. That is clearly in the interests of the drugs industry. But what about nutrition?

Medical students are lucky to get three hours of proper nutrition education. There's lots of money to be made selling drugs. There's none to be made persuading us to eat a more wholesome diet. But we all know nutrition is vital. The World Health Organisation uses birthweight as one criterion for comparing the health of nations.

INTERVIEWER: Change the system, you say?

AUTHOR: Well, we spend fifty billion on what we call the National Health Service. We should really call it the National Disease Service. Our whole approach is cock-eyed. Industry learned years ago that it's daft to spend a fortune on 'quality control' systems in factories, picking up all the faults and then fixing them. Better to spend the money on improving the system so you don't make the faults in the first place. Why can't we do the same with health? The Office of National Statistics tells us there has been a 'relatively large increase' in the number of years a newborn baby can expect to spend in poor health, particularly since 1992. That's outrageous. With all our wealth and all the advances we've been making in medicine, they should be able to look forward to a healthier old age not a *less* healthy one.

INTERVIEWER: Fine, but we can't just stop making chemicals.

AUTHOR: No, but we can use fewer of them on our farms.

INTERVIEWER: And yields would collapse and the price of food would shoot up!

AUTHOR: Yes, there would be lower yields. But it doesn't necessarily follow that prices would have to go up.

INTERVIEWER: Come on! Never heard of the law of supply and demand?

AUTHOR: Indeed I have. And have *you* ever heard of subsidies and how they affect the way that law operates? If we stopped paying billions of pounds to arable farmers to grow food we don't need –

INTERVIEWER: – don't need? That's ridiculous! Of course we

need it. And when you say 'lower' yields why don't you tell us how much lower? If you stopped using chemicals on an arable farm the grain harvest would drop by half. And the farmer would only be able to grow arable crops on two-thirds of the land at any one time because he'd have to rotate his crops.

AUTHOR: But would that really matter? We're already growing so much grain in Europe that we're paying a fortune to arable farmers in 'set-aside' grants. They're actually bribed with our money *not* to grow food. They're also subsidised with arable area payments for what they do grow. And most of the grain ends up getting used for animal feed anyway. So why not have more animals eating the grass and clover that's grown in rotation? The meat will taste better and we can always buy a bit more animal feed from countries like Canada or the United States if we need it.

INTERVIEWER: I'm sure the Chancellor would love that! Why would the government want the country to import any more food than it absolutely has to?

AUTHOR: You have a point. But the more corn the farmers grow, the more they get in subsidies. And who pays for that? We do, the taxpayers. And d'you know how much organic food we're importing? About three-quarters of it. So we would kill two birds with one stone. Surely it makes better economic sense to grow more of our own organic food, not least because organic farms use far more labour than intensive farms. That's another economic benefit. So it creates more jobs and it's kinder to the environment. A double whammy!

INTERVIEWER: This organic food may be fine for well-to-do middle-class people like you and me but it costs a fortune and poorer people can't afford it.

AUTHOR: Yes, it's more expensive in the shops. But that's only a tiny part of the story.

INTERVIEWER: What d'you mean 'in the shops'? That's what matters. That's the whole point.

AUTHOR: No it isn't. Who d'you think pays for the subsidies to grow the grain that we don't need . . . or *not* grow it? We do. Who d'you think pays to clean up the chemical pollution caused by intensive farming? We do. And it costs billions every year. Who d'you think is paying the bill for BSE? That one disaster alone will end up costing close on five billion pounds and the taxpayer is meeting the bill. We pay vast amounts in subsidies to farmers to grow food we don't want in a way that causes great harm to the environment and our health and then we pay another fortune to clean up the mess. It's a brilliant system, isn't it?

INTERVIEWER: Well yes . . . most people agree it's pretty crazy and the European Union's trying to come up with something better but we can't do much until the CAP is reformed.

AUTHOR: Rubbish! Other countries in Europe pay far more to their farmers to convert and run organic operations than we do.

INTERVIEWER: Hah! So after attacking the so-called barley barons for sucking in the subsidies and 'farming for greed and not need' you're now saying we should do the same for organic farmers. What's the difference between subsidising one lot and subsidising the other lot?

AUTHOR: The whole point is that at the moment organic farmers are discriminated against by the subsidy system. If an intensive farmer causes pollution why shouldn't he have to pick up the bill for it? Why should the rest of us pay higher water bills because of agricultural pollution? By paying a bit more to organic farmers at least we'd be subsidising food that people actually want and which would have to be imported otherwise. Ultimately prices in the shops would fall – they're beginning to already – and more people would be able to afford better food. We might even be healthier into the bargain.

INTERVIEWER: You can't prove that.

AUTHOR: Well, I agree that the jury is still out – but there are plenty of pointers. A House of Lords committee reviewed more than a hundred and fifty different investigations comparing organic food with food grown intensively and concluded that organic vegetables tended to contain more vitamins and fewer nitrates.

INTERVIEWER: You're also more likely to get food poisoning from organic food.

AUTHOR: That's complete nonsense.

INTERVIEWER: No it's not. Tesco had to withdraw organic mushrooms from their shops because they were contaminated with E. coli 0157. That's a very dangerous bug. It was a potentially serious incident.

AUTHOR: Yes, it would have been very serious if it had been true. But the Public Health Laboratory Service who carried out the tests admitted five days later that there had been an error in laboratory testing and said, 'There is absolutely no risk to public health from this incident.' Oddly enough, that admission didn't get quite the same coverage in the newspapers as the original story. Wonder why . . .

INTERVIEWER: What about Dennis Avery, the director of food policy at the Hudson Institute in the United States? He's produced a number of reports and books that have been highly critical of organic foods and the risks of food poisoning.

AUTHOR: Indeed he has. He's probably the leading critic of organic farming. One of his books is called *Saving the Planet with Pesticides and Plastic.* Incidentally, the Hudson Institute gets a lot of money from the agrochemical and GM companies. Quite a coincidence, eh?

INTERVIEWER: Doesn't prove he's wrong. What about the

Institute's study based on data from the Federal Centre for Disease Control? It says people are 'eight times more likely to contract the potentially fatal strain of E. coli 0157: H7' from eating organic food than eating conventionally grown food.

AUTHOR: And what about the statement from the Centre which said it's never even carried out such work, let alone produced those results? Even the Food Standards Agency in this country described Avery's 'findings' as nonsense. The United Nations Food and Agricultural Organisation carried out its own research and concluded that organic farming 'potentially reduces the risk of E. coli infection'.

INTERVIEWER: The Institute for Economic Affairs says organic food may well present a danger to children, the elderly and the sick and they should be discouraged from eating it.

AUTHOR: Ah yes, that's the outfit that was founded by the man who made his fortune from setting up the first broiler chicken farm in Britain.

INTERVIEWER: There are other organisations that have attacked organic farming over the years. Are you telling me they're all dodgy?

AUTHOR: No, but I am telling you that even the most independent of research institutes needs money to carry out its work and the agrochemical industry has deep pockets and can be very generous.

INTERVIEWER: And respected scientists can be bought and sold?

AUTHOR: I'm not saying that, but sometimes the connections between people who work for those institutions and the industry can be very close indeed.

INTERVIEWER: If we listened to you we'd end up trusting nobody.

AUTHOR: Probably not a bad philosophy. But how about listening to what the agrochemical industry says? It's been telling us for half a century that its chemicals do no harm to man or beast and the environment actually benefits from intensive farming because the less land you have to farm the less you have to plough up –

INTERVIEWER: Sounds logical.

AUTHOR: Maybe. But time and again we have discovered that one or another pesticide that was supposed to be 'safe' turns out to be anything but. Then it's either withdrawn or – more likely – banned by the authorities.

INTERVIEWER: That's inevitable. We learn as we go along. That's how everything works. And the more experience we have, the fewer risks we run. That's why conventional farming is safer today than it's ever been.

AUTHOR: Quite possibly. But here's what I find really intriguing. Once the industry started investing serious money in genetically modifying plants they had to persuade us that we really needed GM. One of their big arguments was that it would be good for the environment because farmers would need to use less pesticide. Now call me an old cynic again, but if their pesticides are so safe how come it's suddenly so vital that we should use them less? There's a really heart-warming little tale in a Monsanto brochure extolling the virtues of 'Roundup Ready' genetically modified corn. It seems the corn was planted extensively at three Illinois watersheds in 1999. All of those watersheds had had 'chronic problems' in the past meeting water quality standards. But once they started using less pesticide – Bingo! – no more problems. Remarkable, eh? It may have slipped my memory but I can't remember the industry telling us in too much detail how its products polluted the water supply before they wanted to sell GM seed. But then, my memory's not what it was.

INTERVIEWER: You talk about organic farming as though it allows the use of no pesticides at all. But it does.

AUTHOR: Yes, it does. They're made from toxins produced

by the plants themselves as against synthetic chemicals produced in a laboratory.

INTERVIEWER: Makes no difference. They're still poisonous.

AUTHOR: True. But conventional farmers can choose from 2,000 different forms of registered pesticide. Organic farmers have twenty-five and they use them far more specifically and more sparingly. It is extremely rare for organically grown food to contain any residues. They farm in a way that encourages some insects that prey on other insects. They don't just blitz everything. It's a more sustainable approach. Non-selective pesticides kill the pests' natural enemies as well as the pests.

INTERVIEWER: Intensive farming has delivered the goods and you can't get away from that.

AUTHOR: Maybe, though I happen to think they're less than perfect goods. And anyway how much longer can it last? Every year more and more insects and mites become resistant to one pesticide or another. I suppose the manufacturers can keep coming up with new pesticides but is that really a sensible way to be going on?

INTERVIEWER: If the crops keep growing . . . why not?

AUTHOR: That's a big 'if'. There is also increasing evidence that yields are beginning to suffer because of the damage that intensive farming does to the soil.

INTERVIEWER: Precious little evidence.

AUTHOR: That may be because there's precious little research being carried out. And anyway, we've only been farming like this for half a century. That's hardly the blink of an eye in historical terms. Let's see what things look like in another fifty years if we carry on like this. The work they've done in the United States shows that serious amounts of topsoil are disappearing.

INTERVIEWER: Look, we're running out of time –

AUTHOR: – my sentiments precisely –

INTERVIEWER: – I mean for this interview! Isn't it true that you've been a critic of intensive agriculture for a very long time and your views are based on prejudice rather than science? In fact, you simply don't trust scientists and we'd be back in the Stone Age if you had your way.

AUTHOR: It's certainly true that I don't trust ALL scientists. How can anyone when they disagree with each other all the time and when each new generation of scientists disproves half the conclusions reached by the previous generation? One scientist told us in 1988 it was possible for humans to catch a version of mad cow disease. Dozens of other scientists, including some of the most eminent in the land, said he was barking mad. But you're wrong to say I don't believe in science. I do. I just don't define it in quite the way that some modern scientists choose to.

INTERVIEWER: Science is objective and you're anything but.

AUTHOR: No. Science is knowledge and there are all sorts of ways of acquiring knowledge. Of course I'm not a scientist – or even a very good farmer – but I can walk through a field that's rich in herbs and clover and has never seen a bag of chemical fertiliser and I can see cows thriving on it and my intuition tells me that that makes sense.

INTERVIEWER: But the cows won't give as much milk.

AUTHOR: No, thank God! That's something else my intuition tells me. It's not right to force an animal to behave abnormally with an udder that's almost bursting and then kill it after a couple of years when it can no longer 'perform' adequately. It's not right to treat chickens or pigs with cruelty and contempt, as though they're simply units of production.

INTERVIEWER: But if poor people can afford cheap meat –

AUTHOR: But it's *not* cheap. That's the whole point. Some of those 'cheap' chicken breasts are as much as forty per cent water. That's not cheap chicken; it's very expensive water.

But look at the big picture and the other side of the balance sheet. We all pay a huge price in the end for these monstrous practices. Look at BSE. Look at the cost and human misery of the food poisoning from those battery houses. Look at the incalculable cost of antibiotic resistance. Look at the environment. Look at what all those cocktails of pesticide residues may be doing to us. Add up the subsidies. Add up the cost of cleaning up the pollution. How the hell can all that be cheap?

INTERVIEWER: And even if we accept all that do you seriously think that a bit more organic farming will solve everything?

AUTHOR: No, of course not. We need another revolution in our attitudes to food and the way we grow it. Let's stop regarding our land as something that's just there to plunder and our animals as units of productivity for us to exploit. The philosophy of good farming should be the same as the philosophy of good health. Only a fool thinks the NHS should be all about trying to make us better when we fall ill. It should be about promoting good health so that we reduce the risk of falling ill in the first place. And the same with farming. Why treat an animal so appallingly that you have to dose it with chemicals to stop it getting sick? Why treat plants so appallingly you have to spray them endlessly with poisons to kill all the bugs? Why treat the soil as just a lifeless growing medium that you must saturate with man-made chemicals to make the plants grow?

INTERVIEWER: In a nutshell: let's return to the past and to hell with all the advances we've made!

AUTHOR: No. Every day we're learning more about how the world works. We must keep on learning. Organic farming is every bit as much a science as genetic engineering and we need much more knowledge. We know almost nothing about that awesome underworld of tiny creatures that gives the soil its natural fertility. If we spent on that sort of thing one-millionth of what we spend on trying to fight nature one way and another, who knows what the benefits might be? But, yes, we can also learn from the past. My mother used to tell me when I was tiny that a little bit of dirt was good for you. Now the scientists have proved she was right. Not that she needed proof. She had her intuition. It really is the most supreme arrogance to say that we can learn nothing from ten thousand years of agriculture and the only way to do it is the way we've been doing it for the last half-century.

INTERVIEWER: Come on, you're a Luddite!

AUTHOR: No. It is not being a Luddite to say that just because we *can* do something we *should* do it. James Watson made one of the greatest scientific discoveries of our age – the double helix structure of DNA – and he says the belief that science always moves forward is a nonsense. The way farmers cared for their animals was once based on good sense and old-fashioned husbandry. They knew that if you tried to force too much milk from a cow she became ill and if you tried to squeeze too many chickens together in one shed infections spread. There were some basic principles of good husbandry and they stuck to them. Now we are able to flout those principles, knowing that diseases might result, because we also know we can treat those diseases by using masses of antibiotics. It is not just inhumane, it is stupid.

INTERVIEWER: Do you really think we will turn the clock back?

AUTHOR: Look, this is not about turning the clock back. It's about recognising that there is a different way of going forward. And, yes, I do think that will happen.

INTERVIEWER: Just because a few of your friends have persuaded the supermarkets there's money to be made selling organic food to a niche market?

AUTHOR: No. Because millions of people in this country have come to realise the real cost of food – and they're fed up with being told there's only one way of producing it. They want animals to be treated in a civilised and humane fashion. They don't want any more tragedies like BSE. They want to be able to trust their food. They want to know that their children and grandchildren are eating food that is not only safe but nutritious and that the earth in which it is grown will still be capable of growing good food when their children have grandchildren of their own. They know that it's about more than shaving a penny or two off the price of a loaf of bread or squeezing another few litres from an over-bred cow. They know there are ethical and moral issues in producing food and that productivity and profit should not be the only driving forces. They know we don't have all the answers to all the questions. Even the *New Scientist* says it is impossible to foresee what dangers lie in store when we create new life forms artificially as we are doing with genetic engineering. Many people believe that we have been taking a gamble over the last half-century and they are fearful for the next. I think they are right and that's why I wrote this book.

INTERVIEWER: There'll be plenty of big farmers and agricultural industrialists who will say you're naïve.

AUTHOR: You bet there will. Still . . . it won't be the first time I've had an argument, will it?

BIBLIOGRAPHY

Driven by Need

D.A. Thomas, *The Atlantic Star 1939–45*, London, W.H. Allen, 1990

J. Costello & T. Hughes, *The Battle of the Atlantic*, London, Collins, 1977

Press cuttings on the sinking of the *Athenia* and food stock status in the UK, *The Times*, 16 September 1939

S. Broadberry & P. Howlett, *The Economics of World War II, UK: Victory at All Costs*, Cambridge University Press, 2000

P. Howlett, *Fighting with Figures*, London, HMSO, 1995

S. Pollard, *The Development of the British Economy 1914–1980*, London, Edward Arnold, 1983

G. Harvey, *The Killing of the Countryside*, London, Jonathan Cape, 1997

Lloyds War Losses, The Second World War: Vol. 1, British Allied and Neutral Merchant Vessels Sunk or Destroyed by War Causes, London, Lloyd's of London Press, 1989

The U-boat war in the Atlantic 1939–1945, London, HMSO, 1989

Captain S.W. Roskill, *The War at Sea: Vol. 1: The Defensive*, London, HMSO, 1976

D. MacIntyre, *The Battle of the Atlantic*, London, Batsford, 1961

From Caveman to Kitchen

D.C. Sutton, *The History of Food*, Coventry, Chapelfields, 1982

F.C. Accum, *There is Death in the Pot, A Treatise on the Adulteration of Foods and Culinary Poisons*, London, 1820

E.V. McCollum, *A History of Nutrition*, Boston, Houghton Mifflin, 1957

D.J. Oddy & D.S. Miller (eds), *The Making of the Modern Diet*, London, Croom Helm, 1976

H.D. Renner, *The Origin of Food Habits*, London, Faber & Faber, 1944

J. Wynne-Tyson, *Food for a Future: the ecological priority of a humane diet*, London, Davis-Poynter, 1975

A.H. Halsall, *Food and its Adulterations, Comprising the Reports of the Analytical Sanitary Commission of 'The Lancet' for the years 1851–1854*, London, 1855

Select Committee Reports on Food Adulteration, 3rd Report (HC 379) VIII, 1, 1856

J. Mitchell, *Treatise on the falsifications of food and the chemical means employed to detect them*, London, 1848

J.C. Drummond, 'An 18th Century Experiment in Nutrition', *The Lancet*, Vol. 229, pp.459–462, 1935

A.W. Flux (Journal of Royal Statistics Society), 'Our Food Supply Before and After The War', Vol. 93, pp.538–56, 1930

D.J. Barker, *Fetal and infant origins of adult disease*, British Medical Journal, 1992

W.P. James, G.G. Duthie & K.W. Wahle, 'The Mediterranean Diet: Protective or Simply Non Toxic', *European Journal of Clinical Nutrition*, Vol. 43, 1989

J.M. Slater (ed), *Fifty Years of the National Food Survey 1940–1990*, London, HMSO, 1991

J. Boyd-Orr, *Food, Health and Income*, London, Macmillan, 1936

D.J. Oddy & D.S. Miller, *Diet and Health in Modern Britain*, London, Croom Helm, 1985

J. Boyd-Orr and D. Lubbock, *Feeding the People in War-time*, London, Macmillan, 1940

J.A Paris, *A Treatise on Diet*, London, 1826
Andrew Wynter et al., various articles, *The London Review*, 1850
M. Leeming, *History of Food from Manna to Microwave*, London, BBC, 1991
Hernes, G., 'Eating to Your Heart's Delight', *Science of the Total Environment*, Vol. 249, No. 1–3, 2000
E.J.T. Collins, 'The Consumer Revolution and the Growth of Factory Foods', in *The Making of the Modern British Diet*, Croom Helm, 1976
A. Fenton and E. Kisbán (eds), *Food in Change, Eating Habits from the Middle Ages to the Present Day*, Edinburgh, Donald in association with the National Museums of Scotland, 1986
J. Morris and R. Bate, *Fearing Food: Risk, Health and Environment*, Oxford, Butterworth-Heinemann, 1999
Erik Millstone, *Food Additives*, Harmondsworth, Penguin, 1986

These Toxic Times

A. Watterson, *Pesticide Users' Health and Safety Handbook*, Aldershot, Gower Technical, 1988
D. Pimentel & H. Lehman, *The Pesticide Question: Environment, Economics and Ethics*, London, Chapman & Hall, 1993
The BMA Guide to Pesticides Chemicals and Health, London, Edward Arnold, 1992
P. Beaumont, *Pesticides, Policies and People*, London, Pesticide Action Network UK, 1993
Barbara Dinham, *The Pesticide Hazard*, Pesticide Action Network UK, 1993
Public Health Impact of Pesticides Used in Agriculture, World Health Organisation, 1990
L.H. Campbell & A.S. Cooke, *The Indirect Effects of Pesticides on Birds*, Peterborough, Joint Nature Conservation Committee, 1997
M.Allsopp, B.Erry, R.Stringer, P.Johnston, D.Santillo, *A Recipe for Disaster: A Review of Persistent Organic Pollutants in Food*, London, Greenpeace, 2000

M. Allsopp, R. Stringer & P. Johnston, *Unseen Poisons: Levels of Organochlorine Chemicals in Human Tissues*, London, Greenpeace, 1998

M. Allsopp, R. Stringer, P. Johnston, D. Santillo, *The Tip of the Iceberg: State of Knowledge on Persistent Organic Pollutants in Europe and the Arctic*, London, Greenpeace, 1999

World Cancer Research Fund in association with The American Institute for Cancer Research, *Food Nutrition and the Prevention of Cancer: A Global Perspective*, London, WCRF, 1997

The Game Conservancy Trust, *Lowland Agriculture Into the 21st Century*, Fordingbridge, Game Conservancy, 2000

Nutritional Aspects of the Development of Cancer, Report of the Working Group on Diet and Cancer of the Committee on Medical Aspects of Food and Nutrition Policy, London, The Stationery Office, 1998

Ministry of Agriculture, Fisheries & Food Pesticide Safety Directorate Health and Safety Executive, Annual Report of the Working Party on Pesticide Residues 1998, 1999 & the quarterly monitoring supplements in 2000

A.P Høyer, P. Grandjean , T. Jørgensen, J.W. Brock, H.B Hartvig, 'Organochlorine exposure and breast cancer', *The Lancet*, Vol. 352, pp.1816–20, 1998

C. V. Howard and G. Staats de Yanes, 'A Hazardous Risk?', *Environmental Health*, issue 7/8, pp.12–14, 1998

'Food safety in the 21st century', *Dairy, Food and Environmental Sanitation*, Vol. 19, No.11, 1999

S. Valentine, 'Food and nutrition in the twenty-first century curriculum', *Nutrition & Food Science*, Vol. 30, No. 3, 2000

Saving Lives: Our Healthier Nation, London, The Stationery Office, 1999

H. Tent, 'Research on food safety in the 21st Century', *Food Control*, 10, 1999

L.M. Bush and R.A. Williams, 'Diet and health: new problems/new solutions', *Food Policy*, 24, 1999

P. Conford, 'A testament for youth', *Living Earth*, No. 202, Apr–Jun, 1999
M. Kid, 'Food safety: consumer concerns', *Nutrition & Food Science*, Vol. 30, No. 2, 2000
M. O'Keeffe & O. Kennedy, 'Residues: a food safety problem?', *Journal of Food Safety*, Vol. 18, No. 4, 1998
Department of Health, Committee on Medical Aspects of Food and Nutrition Policy (COMA), Seventh Annual Report, 1999
M. Rayner, 'The truth about the British diet'; 'We are becoming healthier eaters, government ministers claim – The reality is rather more worrying'; 'We are dining out on the fat of the land', *New Scientist*, London, July 1989
'Hormone-disrupting chemicals found in baby food', Friends of the Earth Press Release, 1 December 2000
UK Pesticide Guide, British Crop Protection Council, Farnham, 1999
L.E. Gray, J. E. Monosson and W.R. Kelce, 'Environmental antiandrogens: low doses of the fungicide vinclozolin alter sexual differentiation of the male rat', *Toxicology and Industrial Health*, Vol. 15, pp.48–64, 1999
A. Mantovani, F. Maranghi, C. Ricciardi, C. Macri, A.V. Stazi, L. Attias and G.A. Zapponi, 'Developmental toxicity of carbendazim: comparison of no-observed-adverse-effect level and benchmark dose approach', *Food and Chemical Toxicology*, Vol. 36, pp.37–45, 1998
Pesticide Action Network briefing on aldicarb
Pesticide Action Network briefing on lindane
Strang-Cornell Cancer Research Laboratory, New York, briefings provided on lindane, carbendazim, chlorpyrifos and vinclozolin, 2000
A. White, 'Children, pesticides and cancer', *Ecologist*, 28, Mar/Apr 1998
Residues of some veterinary drugs in animals and foods, prepared by the fiftieth meeting of the Joint FAO/WHO Expert Committee on Food Additives, Rome, FAO, Feb 1999

J. Erlichman, *Gluttons for Punishment*, Harmondsworth, Penguin, 1986

R. Beaglehole & R. Bonita, 'Public Health at the Crossroads', *The Lancet*, Vol. 351, February 1998

J.G Morris, 'Current trends in human diseases associated with foods of animal origin', *Journal of American Veterinary Medical Association*, 15 December 1996

E. Carlsen, A. Giwereman, N. Keiding and N.E. Skakkebaek, 'Evidence for decreasing quality of semen during the past 50 years', *British Medical Journal*, 305 pp.609–12, 1992

IEH (1995), *IEH Assessment on Environmental Oestrogens: Consequences to Human Health and Wildlife*, (Assessment A1), Leicester, UK, MRC Institute for Environment and Health, 1995

'Scientists link falling sperm to chemicals in food', *The Sunday Times*, 5 May 1995

A. Chowdhury, H. Venkatakrishna-Bhatt and A. Gautum, 'Testicular changes of rats under lindane treatment', *Bulletin Env. Cont. Toxicol. Appl. Pharmacol.*, Vol. 90, pp.330–6, 1987

A. Abell, E. Ernst and J.P. Bonde, 'High sperm density among members of organic farmers' association', *The Lancet*, Vol. 343, No. 8911, p.1498, 1994

The World Beneath Our Feet

T. Brock and M. Madigan, *Biology of Micro-organisms*, Englewood Cliffs, Prentice-Hall, 1988

B.A. Croft, *Arthropod Biological Control Agents & Pesticides*, New York, Wiley, 1990

D. Pimentel & H. Lehman, *The Pesticide Question: Environment, Economics and Ethics*, London, Chapman & Hall, 1993

C.A. Edwards, 'Impact of Herbicides on Soil Ecosystems: The Importance of Integration in Sustainable Agricultural Systems', *CRC Critical Reviews in Plant Sciences*, 8:3, pp.221–57, 1989

MAFF *Code of Good Agricultural Practice*, London, The Stationery Office, 1998

Daniel Hillel, *Out of the Earth*, New York, Free Press, 1991
'Sustainable Use of Soil', 19th Report of the Royal Commission on Environmental Pollution, Cm. 3165, London, 1996
Neil Fuller, *Blueprint for Eco-Farming*, lecture given at 1997 Acres USA conference, St Louis, MO, 4–6 December 1997
A Blueprint for Sustainable Soil Management (A submission to the Royal Commission on Environmental Pollution by the Soil Association), February 1995
Margaret P. Rayman, 'The Importance of Selenium to Health', *The Lancet*, 15 July 2000
D. Hillel, *Out of the Earth: Civilization and the Life of the Soil*, Berkeley, University of California Press, 1992

Fear of Fish

Lochaber & District Fisheries Trust, Annual Report 1998–1999
Anouk Ride, 'Dead in the Water', *New Internationalist*, July 2000
Michael Wigan, 'Farmed Salmon Kill the Rivers', *The Field*, June 2000
Steve Farrar, *The Sunday Times*, June 1999; Nick Nuttall, *The Times*, April 2000; Rob Edwards, *Sunday Herald*, February 2000
Don Staniford, 'The One That Got Away', Friends of the Earth, Edinburgh, Scotland, 2000
R.L. Naylor, R.J. Goldburg, J.H. Primavera, N. Kautsky, M.C.M. Beveridge, J. Clay, C. Folke, J. Lubchenko, H. Mooney & M. Troell, 'Effect of aquaculture on world fish supplies', *Nature*, 29 June 2000
L.J. Forristal, *The World & I*, 'Is Something Fishy Going On?', May 2000
R. Edwards, 'Infested waters: sea lice from salmon farms threaten Scotland's sea trout', *New Scientist*, 4 July 1998
Written testimony made to Friends of the Earth Scotland by Jackie Mackenzie, Edinburgh, April 2000
Written testimony made to Friends of the Earth Scotland by Johnny Parry, Edinburgh, April 2000

Department of Environment, Transport and the Regions, Report: *Sustainable Production & Use of Chemicals*, 1999

Working Party on Pesticide Residues, Annual Report, September 1999

Scottish Environment Protection Agency, Annual Report, 1999

D.W. McKay, 'Perspectives on the Environmental Effects of Aquaculture', lecture given at Aquaculture Europe '99 conference, Trondheim, 7–10 August

Veterinary Medicines Directorate Annual Report on Surveillance for Veterinary Residues, July 1999

Food Contaminants Division, Food surveillance information sheet 184, 'Dioxins and PCBs in UK and imported marine fish', Food Standards Agency, August 1999

B. Moore, N. Boyns, and H. Tilley, *The Economic Impact of Salmon Farming*, Edinburgh, Scottish Executive, 1999

Food Safety Issues Associated with Products from Aquaculture, World Health Organisation, 1999

Battling with Bugs

Resistance to Antibiotics and Other Antimicrobial Agents, Evidence submitted to the House of Lords Select Committee on Science and Technology, London, The Stationery Office, 1998

Government Response to the House of Lords Select Committee on Science and Technology Report, *Resistance to Antibiotics and other Antimicrobial Agents*, London, The Stationery Office, 1998

Advisory Committee on Microbiological Safety of Food, *Report on Microbial Antibiotic Resistance in Relation to Food Safety: Synopsis*, London, The Stationery Office, 1999

World Health Organisation, *The Medical Impact of the Use of Antimicrobials in Food Animals*, report of a WHO meeting, Berlin, 1997

M.H. Richmond, 'Why has Swann failed?', *British Medical Journal*, Vol. 280, No. 6225, pp.1195–6, 1980

M.H. Richmond, 'The emergence of antibiotic resistance in bacteria and its implications for antibiotic use', *Ten Years on from Swann*, Association of Veterinarians in Industry Symposium, 1981

R.J. Cook, 'Antimicrobial resistance: use in veterinary and human medicine', *Journal of Antimicrobial Chemotherapy*. Vol. 9, p.435, 1997

A Force for Change, White Paper on the Food Standards Agency, HMSO, January 1998

I. Hunt, 'Antibiotics: agents lose their cutting edge', *Guardian*, 5 May 1998

Q.A. McKellar, 'Antibiotics and resistance in farm animals', *Nutrition & Food Science*, No. 4, July/August 1999

P. Brown and J. Meikle, 'Alarms rang 50 years ago', *Guardian*, 7 September 1999

S. Levy, *The Antibiotic Paradox: How Miracle Drugs are Destroying the Miracle*, London, Plenum Publishing, 1992

'Less is more', *New Scientist*, 25 April 1998

'Resistance to antibiotics – a threat to public health', opinion of the European Commission Consumer Committee, adopted 1 March 1999

'Antibiotic use in food-producing animals must be curtailed to prevent increased resistance in humans', WHO Press Release, October 1997

N. Frimodt-Moller, N. Rosdahl and H. Caspar Wegener, 'Microbiological resistance promoted by misuse of antibiotics: a public health concern', *European Journal of Public Health*, No. 8, pp.193–4, 1998

'Resistance to antibiotics as a threat to Public Health', opinion of the European Community's Economic and Social Committee, September 1997

Antimicrobial Feed Additives, report of the Commission on Antimicrobial Feed Additives, Stockholm, 1997

Swedish Ministry of Agriculture, Food and Fisheries, *Can we use less antibiotics?: A brochure on antibiotics in animal feed and how they affect humans and animals*, Stockholm, 1997

J. Meikle, 'Farm antibiotics pose risk to human health', *Guardian*, 19 August 1999

J. Meikle, 'Farmers promise to cut back antibiotics', *Guardian*, 29 June 1999

Diseases Fighting Back, London, Parliamentary Office of Science and Technology, October 1994

T.H Jukes, 'Effects of low levels of antibiotics in livestock feeds', *Agricultural Uses of Antibiotics*, ACS Symposium Series, New York, American Chemical Society

Joint Committee on the use of antibiotics in animal husbandry and veterinary practice, Cm. 4190, London, 2000

'Veterinary Residues in Animal Products 1986–1990', MAFF Food Surveillance Paper No. 33, London, HMSO, 1992

J. Bates, J. Zoe Jordens and D.T. Griffiths, 'Farm animals as a putative reservoir for vancomycin-resistant enterococcal infection in man', *Journal of Antimicrobial Chemotherapy*, Vol. 34, No. 4, pp.507–516

The New Gene Genie

'The Promise of Plant Biotechnology', a Monsanto Biotech Information Leaflet, Monsanto US, St Louis, 2000

Robert B. Shapiro, Chairman and CEO Monsanto Corporation, letter to Gordon Conway, President Rockefeller Foundation, October 1999

Professor Philip James, Director, Rowett Research Institute, Evidence to House of Lords, January 1999

Dr Mae-Wan Ho (Journal of 'Horizontal gene transfer, hidden hazards of genetic engineering', November 2000) *Institute of Science in Society*

Paul R. Billings & others, 'Human Germline Gene Modification: A Descent', *The Lancet*, Vol. 353, No. 9167, May 1999

D. MacKenzie, 'Unpalatable Truth', *New Scientist*, Vol. 162, No. 2182, April 1999

S. Dibb and S. Mayer, 'Biotech – The Next Generation', The Food Commission, April 2000.

T. Traavik, *Too early may be too late*, University of Tromsø, 1999
Judith C. Juskevich and Greg C. Guyer, 'Bovine growth hormone: human food safety evaluation', *Science*, Vol. 249, August 1990
Anne McIlroy, article in *Toronto Globe and Mail*, September 1998
Expert Group on GM Food, Report, London, Medical Research Council, June 2000
The Impact of Genetic Modification on Agriculture, Food and Health, London, British Medical Association Board of Science and Education, May 1999

If I May Just Finish

Healthy Life Expectancy in Great Britain 1980–96, Sue Kelly & Allan Baker, Office of National Statistics & Department of Health

INDEX